28-

PENGUIN BOOKS

# NEW APPLICATIONS OF MATHEMATICS

Christine Bondi graduated in mathematics and proceeded to a
Ph.D., both at Newnham College, Cambridge. In 1947 she
married Hermann Bondi, until recently Master of Churchill
College, Cambridge. After bringing up a family of five (all of
whom passed A-level mathematics, though their other subjects
spanned a wide range) she went into mathematics teaching and
taught in secondary schools in Surrey, and then in a sixth-form
college for many years. In 1985–6 she chaired a working party
which produced a Royal Society/Institute of Mathematics
report on the subject of girls and mathematics. She believes that
mathematics is a subject for all, and that it should be taught so
that its applications come alive. This book reflects that interest.

D0067517

# NEW APPLICATIONS OF MATHEMATICS

EDITED BY
CHRISTINE BONDI

PENGUIN BOOKS

PENGUIN BOOKS

Published by the Penguin Group
27 Wrights Lane, London W8 5TZ, England
Penguin Books USA Inc., 375 Hudson Street, New York, New York 10014, USA
Penguin Books Australia Ltd, Ringwood, Victoria, Australia
Penguin Books Canada Ltd, 2801 John Street, Markham, Ontario, Canada L3R 1B4
Penguin Books (NZ) Ltd, 182–190 Wairau Road, Auckland 10, New Zealand

Penguin Books Ltd, Registered Offices: Harmondsworth, Middlesex, England

First published 1991
1 3 5 7 9 10 8 6 4 2

Filmset in Monotype Lasercomp Times by
The Alden Press, Oxford

Made and printed in Great Britain by
Richard Clay Ltd
Bungay, Suffolk

# CONTENTS

## CONTENTS

# FOREWORD

The Institute of Mathematics and its Applications (IMA) was formed in 1964 to encourage and develop the applications of mathematics in all relevant fields of intellectual endeavour – educational, industrial, commercial, medical, cultural, and so on. It has grown to become the largest representative body for professional mathematicians in the world, with a vigorous and expanding programme of activities which reflects the ever increasing potential for developing and applying mathematics in today's world, and tomorrow's.

The IMA has been responsible for getting this book together, and records its thanks to all the authors who have contributed. They, in their different ways, are expressing the fun and pleasure that can be found in using mathematics, and describing the often surprising places where mathematics is used. Dr Christine Bondi, who has edited the book, deserves special thanks. By patiently helping and encouraging the authors she has produced a coherent, readable volume that is also informative and useful. David Nelson has contributed at every stage of the book's development and his help, work and imagination are most gratefully acknowledged. Special thanks, too, are given to my colleagues Norman Clarke, who originally suggested the book, and Hilary Brown, who made it happen.

Catherine Richards
Secretary and Registrar
The Institute of Mathematics and its Applications
16 Nelson Street
Southend-on-Sea
Essex SS1 1EF

# PREFACE

Few teachers of A-level mathematics will be unfamiliar with the plaintive question, 'What is the use of this topic?'

The intention of this book is to provide some answers. At a time when there are changes of all sorts taking place in school education, with the introduction of GCSE, the greater emphasis on project work and a greater demand for relevance, we hope this intention will have been fulfilled, both by providing long-term interest and motivation and by supplying immediate ideas for worthwhile projects.

Mathematics is by no means an exclusive and purely academic subject: at one level or another it is a subject for all, and the awareness of this is increasing from nursery school onwards. This book is aimed at A-level (or equivalent) students and their teachers. Many of those students are now much more broadly based in their interests than students of the past, and the applications of mathematics are much wider. The majority of students studying mathematics post-GCSE are interested in pursuing those applications rather than in becoming pure mathematicians; they are the people we are catering for here. That is not to say that the elegance and fascination of apparently 'pure' mathematics do not have their place in mathematical education – far from it: two chapters (Chapters 8 and 11) in this book describe an application of a topic that was originally believed to be without possible use. But for the most part it is the visibly applicable aspects that this book is tackling.

It is not necessary to read this book straight through. There is some deliberate ordering of chapters: for example, it would be a mistake to read Chapter 4 before at least part of Chapter 1, but since the mathematics generally becomes rather more difficult as a chapter proceeds, students may well find it better to leave the later parts of some chapters to a second reading, as their courses and experience develop. In any case, the earlier parts of each chapter contain descriptions of the problems that can be investigated with the relevant mathematical topics, and in some chapters there follows an explanation in textbook style. While there has been no intention to make this a textbook, it seemed appropriate to include such material here, both to ensure that the mathematical ideas had been introduced and to emphasize their relevance.

What is important is not so much that the mathematics should all be understood in detail immediately, as that the student should see something of the problem that can be investigated or even solved. We hope it will be an enjoyable as well as an instructive experience.

Christine Bondi
Cambridge

# 1

## USING FUNCTIONS AND GRAPHS
## TO UNDERSTAND OIL WELLS
## AND BUSINESS ACCOUNTS

### JULIAN HUNT

**Two scenes from real life**

*Scene 1*

In a small control cabin next to an oil rig, two engineers, Izzy Husain and Lee Brown, are looking at two graphs. They are discussing the 'log' of the 'run' they have just made by dropping an instrument 800 m down the oil well. The results were recorded by the small minicomputer and the output was displayed as graphs. The first graph (Fig. 1.1(a)) shows the rate of flow of oil at different depths down the well. (They call the rate of flow by the symbol $F$; it is measured in cubic metres per second. They call the depth of each measurement by the symbol $x$; it is measured in metres.) The second graph (Fig. 1.1(b)) shows the density $D$ of the mixture of oil, water and gas in the well at different depths ($D$ being measured in kilograms per cubic metre).

IZZY: This log seems generally OK but there are a few funny things about it. From the bottom of the well up to a depth of 600 m, $F$ is increasing; something is coming into the well all right. The density graph shows there's water at the well bottom and that $D$ drops just when $F$ starts to increase. So there must be oil coming in there. But what do you make of $F$ suddenly dropping at 600 m, Lee?

LEE: Well, this must be a classic case of oil *leaving* the well and entering the rock formation. The only way $F$ can drop is if there is a flow *out* of the well. It might be an idea to see how much $F$ changes for each metre up the pipe. [Lee gets the computer to calculate the *change* in $F$ between 600 and 601 m, and then 601 and 602 m, and so on. She calls this $q$, and the computer outputs a

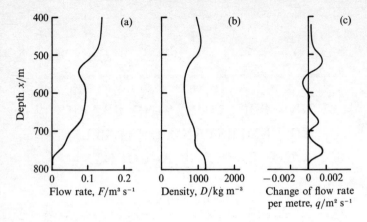

*Figure 1.1.* Graphs of (a) flow rate, (b) density and
(c) change of flow rate from the oil well log, discussed by
Izzy and Lee in Scene 1.

graph (Fig. 1.1(c)) of $q(x)$ for each metre down the well.] Izzy, this
formation is even worse than I thought. It is only really producing
between 650 and 700 m, and between 720 and 780 m. Between 550
and 600 m, $q$ is *negative*, and then between 500 and 550 m $q$ is quite
positive again.

IZZY: But don't you see, Lee, that last blip on your $q$ curve just
below 500 m is not oil entering – it looks like water, because the
density curve $D$ increases there. So we'll have to put in our report
that this formation is really useless above a depth of 650 m and
above that they should block up the perforations.*

LEE: I agree, but we ought to compare these results with the output of the computer model for the formation. Let's call it up on
the monitor.

[Lee keys in a few instructions, and a prediction curve of density
(not shown in Fig. 1.1) comes up on the screen. Before blasting
holes in the walls of the steel pipe, they had run two tests to
measure how the rock, oil and water around the well had absorbed
radiation in the form of rays emitted from a nuclear radioactive
source on their measuring instrument, and to measure how the
electrical resistance of the liquids varied with the rock found
around the wall at different depths. They used these measurements, together with complex computations of the scientific laws
governing the absorption of radiation and the distribution of

* Their term for the holes in the pipe where the oil enters.

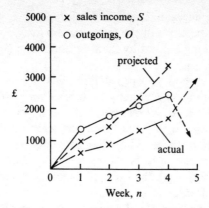

*Figure 1.2.* Graph showing sales and outgoings for the JANSE business, discussed by Tam, Jackie and Hal in Scene 2.

electrical current, to assess the amount of oil and water in the formation.]

The computer model shows the right distribution of oil and water, but it does not predict that oil might flow out of the pipe! It is interesting that it does seem to suggest that the maximum production might come from different layers, 650 and 750 m down, where we find $q$ is a maximum.

### Scene 2

Round a table at the back of JANSE (a small business recently set up to design and sell a new style of denim clothing), the owners, Jackie Campbell and Hal Black, with an accountant from their bank, Tam Lubbock, are looking at the figures of the business.

TAM: Well, you've been going for a month now, but the sales aren't picking up as fast as you projected, and the outgoings on staff, rent and so on are steadily eating up your initial capital. I'd like to hear what your plan is over the next few months or so.

JACKIE: It's true that we thought April and May would be good for people to buy their summer jeans, but it was either pelting with rain or baking hot. Either they didn't shop at all, or else it was too hot for denim. I think . . .

HAL: Stop it! I can't stand all this gloomy waffle. Let's take the actual figures, and the figures we *thought* we'd get, and draw two plots of our income, and then a plot of our outgoings. [See Fig. 1.2.] Let the crosses be our income, and the circles our outgoings. When we started, we hoped to be on the projected curve. The projected curve crosses the curve of the outgoings by week 3, so we should have been making a profit by then. But our actual curve is below the outgoings.

TAM: So there are now two possibilities, which I have marked on the graph: you must get either your income up or your outgoings down. Which is it to be? Of course, if you do neither . . .

JACKIE: I refuse even to think of cutting back on our staff after all they have done for us. I know we can jump our sales up to the planned curve. The word is getting about, we have got fantastic new designs and, thank the Lord, the weather has returned to the usual British drizzle. So . . .

[So, she scratched out the downward arrow and underlined the upward arrow – let's hope she was right!]

## Functions tell us about trends as well as numbers

Scenes like these happen all the time, in industry, in small businesses, in government offices and on school committees, as well as in more exotic locations like spacecraft (though the data and curves would more likely be on a screen). They show one important aspect of how mathematics really works in the everyday world, and also in the most remote places and unusual activities, by providing ways of expressing and working out complicated problems and situations. Much of the dialogue in our scenes is about relations between different quantities: about trends up or down, or sudden changes, or whether the sign was positive or negative. We have been looking at functions in action.

Functions are used in mathematics to give a precise description and analysis of these kinds of graph, or to describe relations between different 'quantities' or different trends. This is just as much a part of mathematics as dealing with numbers. In other words, mathematics helps with 'qualitative' as well as 'quantitative' analysis.

Of course, the qualitative way in which a mathematician looks at the world is quite different from how a poet or artist sees it. The great advantage of knowing some mathematics is that one can see something of the beauty of both kinds of vision.

The Old Testament poet described how individuals and nations enter into conflicts:

And thine eye shall not pity; but life shall go for life, eye for eye, tooth for tooth, hand for hand, foot for foot. (Deuteronomy 19:21)

When thou comest nigh unto a city to fight against it, then proclaim peace unto it . . . And if it will make no peace with thee, but will make war against thee, then thou shalt besiege it. (Deuteronomy 20:10–12)

Compare these with a mathematical model for how people and nations get into fights with one another. In 1919 L. F. Richardson (in *The Psychological Causes of War*) suggested some equations for how the 'animosity' or 'aggressive feelings' of two sides, the Reds and the Blues, grow. If the animosities are represented as $A_R$ and $A_B$, the equations are

$$\frac{dA_R}{dt} = kA_B, \qquad \frac{dA_B}{dt} = kA_R \qquad (1.1)$$

These equations state that the animosity of the Reds grows at a rate, $dA_R/dt$, that is proportional to the animosity of the Blues, and vice versa (i.e. if the Blues get more aggressive, so do the Reds!).

Or compare how Leonardo da Vinci drew vortices (see Fig. 1.3) with the mathematical laws of vortex motion developed by Lord Kelvin and Hermann von Helmholtz in the nineteenth century:

$$\frac{D}{Dt}\omega = (\omega . \nabla)u, \qquad \omega = \nabla \wedge u, \qquad \nabla . u = 0 \qquad (1.2)$$

These equations* describe the flow in terms of its velocity $u$ and the swirling motions in the vortices ($\omega$ is the 'vorticity'). They show that the vortices keep swirling even as they move around the flow, the vortices have no ends within the flow, and the little vortices twist and stretch each other, untangling only rather slowly. Much of this is depicted in Leonardo's sketches, and it is all represented by equations (1.2).

In this chapter we introduce the idea of functions. We consider them in their simplest form, and then see how the simple forms can be used as building blocks for more complicated functions, such as the ones in our two scenes. To understand how functions are *used*, we must also consider the relations *between* functions, as Izzy and Lee were discussing, and where such relations come from – whether from data in a particular situation or from general scientific laws. Compared with just ten years ago, functions are now being used in different ways. In this chapter and later ones we'll try to hint at the new developments, which are not only exciting and interesting but useful, too.

---

* The precise meaning of these equations does not concern us here, but if you are interested look in a book on fluid dynamics such as M. J. Lighthill's *An Introduction to Theoretical Fluid Mechanics* (Oxford University Press, 1986).

*Figure 1.3.* Leonardo da Vinci's sketches of fluid flow and vortices, reproduced by gracious permission of Her Majesty The Queen. Equations can provide us with another kind of picture.

### Defining functions and variables and watching out for apples and pigs

The graphs in Figs. 1.1 and 1.2 illustrate some of the main features of functions. A *function* is the mathematical term for describing how one *variable* is related to another. In the simplest kind of function, different *values* of one quantity are *related to* (or *correspond to*) the values of another quantity. In our first example, the flow rate $F$ in the pipe, in cubic metres per second, is related to the depth $x$ down the oil well, in metres; in the second example the sales $S$, in £, are defined for the week in which they occurred, $n$.

Such relations can be expressed as graphs, as we have seen, or as lists of numbers (see Table 1.1). In mathematical notation, we express such relations as

$$F = f(x) \text{ or } x \mapsto F \text{ in our first example, Fig. 1.1 (1.3)}$$

$$S = f(n) \text{ or } n \mapsto S \text{ in our second example, Fig. 1.2 (1.4)}$$

The commonest symbols are

$$y = f(x)$$

The arrow notation indicates that for each specified value of $x$, we can define a value of $y$.

If functions were no more than a means to draw graphs from a set of numbers, or express the relations by symbols, there would be no more to say about them. In fact, though, there are general rules and ideas about functions, much like general rules about numbers. The ideas of number apply to any collection of similar objects. (So two apples plus two apples equals four apples, and the same applies to pears or pigs.) The basic elements of number are the integers, from which we go on to build fractions, irrational numbers, and so on. In the same way there are an infinite number of functions, ranging from the simplest type to ones of limitless complexity.

*Table 1.1.* The data plotted in Fig. 1.2.

| Week, $n$ | Sales income (actual), $S$ | Sales (projected) | Outgoings (actual) |
|---|---|---|---|
| 1 | 600 | 1000 | 1400 |
| 2 | 700 | 1400 | 1700 |
| 3 | 1300 | 2400 | 2100 |
| 4 | 1700 | 3400 | 2400 |

Mathematical analysis is simplest and most general when it is dealing with objects or quantities of the same sort; for shorthand, we say we are dealing with *pure numbers*. But what would happen if we wanted to add two pigs to two apples? We would then obviously need a set of rules different from those for adding just pigs or just apples, and would have to introduce a new notation, such as $(2, 2)$, and that leads on to vectors and matrices (see Chapter 2).

In general, then, we are thinking about pure numbers, and we use the standard symbols for a function $y = f(x)$, or $x \mapsto y$. In addition, it is always necessary to consider the *set* of values of $x$ (or *interval*) for which the function is defined, which we denote by $A < x < B$; that is, $x$ lies between the values $A$ and $B$ (either or both of which could be negative numbers).

It is also necessary to consider whether $y$ is defined for every value of $x$ on this interval, or merely at specific points along it – that is, whether the graph of the function is like Fig. 1.1 or Fig. 1.2. The first is a smooth function, and the second is defined only for complete weeks. (It is common practice in the latter case to draw straight lines between the points on the graph.)

### Building blocks for functions and making approximations

The simplest examples of functions are listed below and sketched in Fig. 1.4:

$$y = b, \qquad \text{where } b \text{ is constant: a straight line}$$
$$\text{parallel to } y = 0 \text{ (the } x \text{ axis)} \qquad (1.5)$$

$$y = x, \qquad \text{a straight line at } 45° \text{ to } y = 0 \qquad (1.6)$$

$$y = ax + b, \quad \text{where } a \text{ and } b \text{ are constant: a straight}$$
$$\text{line of slope } a \text{ passing through the}$$
$$\text{point } (0, b) \qquad (1.7)$$

These are examples of *linear relations* because their graphs are straight lines. They are also functions, since for a given value of $x$ there is just one value of $y$.

An important property of straight lines is that they have the same slope at each point along them. The slope is the ratio of the vertical distance $v$ to the horizontal distance $h$ in going between two points $P_1$ and $P_2$ on the line (see Fig. 1.4). For the points on the line $y = x$, we see that $v = \frac{1}{2}$ and $h = \frac{1}{2}$, so the slope is 1. For the line $y = 2x + \frac{3}{2}, v = 1$ and $h = \frac{1}{2}$, so the slope is 2. For the general line

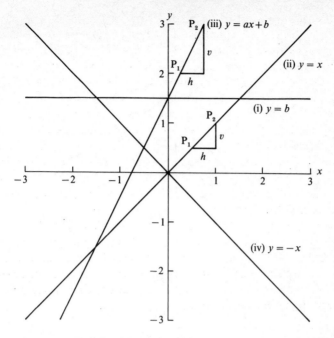

*Figure 1.4.* Building blocks for functions: the simplest, linear (straight line) graphs. (i) $y = b$ (with $b = \frac{3}{2}$), (ii) $y = x$, (iii) $y = ax + b$ (with $a = 2$ and $b = \frac{3}{2}$) and (iv) $y = -x$.

given by equation (1.7), the slope is $a$. For the line $y = -x$, $y$ decreases by $\frac{1}{2}$ when $x$ increases by $\frac{1}{2}$, so the slope is $-1$.

More complex *non-linear functions*, leading to curves, can be obtained when, for each value of $x$, the corresponding value of $y$ is given by $x$ multiplied by itself once ($x^2$), or twice ($x^3$):

$$y = x \times x = x^2, \qquad \text{parabola} \qquad (1.8)$$

$$y = x \times x \times x = x^3, \quad \text{cubic} \qquad (1.9)$$

and so on (see Fig. 1.5). All the functions in equations (1.5)–(1.7) and (1.9) are derived for the whole range of $x$, from $-\infty$ to $+\infty$, where the symbol $\infty$ stands for infinity. But in equation (1.8) the range of values of $y$ extends only from 0 to $\infty$.

We now have enough information to understand how *approximate* curves and functions are built up when only a few points of the function have been defined. Consider, for example, a small hill that is to be removed by bulldozers for a new road, for which only a few measurements have been made of the height of the hill (above a

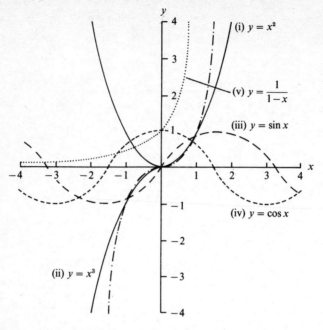

*Figure 1.5*. Building blocks for functions: some basic
non-linear functions. (i) Parabola, $y = x^2$, (ii) cubic, $y = x^3$,
(iii) sine curve, $y = \sin x$, (iv) cosine curve, $y = \cos x$,
and (v) hyperbola, $y = 1/(1 - x)$.

known level) at different points. The shape of the hill needs to
be defined in order to calculate the volume of earth to be removed.
One approach would be to put straight lines between the points on
a graph, as in Fig. 1.6(a), i.e. to use equation (1.7) as the building block, with different values of $a$ and $b$ between each pair of
points. But the resulting function for the hill surface is unrealistically
jagged – hills don't look like that. A more realistic, but more complex, approach would be to use a combination of a cubic, a parabola
and straight lines. If we add all the curves (1.7), (1.8) and (1.9)
together, with constants $c$ multiplying $x^2$ and $d$ multiplying $x^3$, we get
a general cubic curve whose equation is

$$y = a + bx + cx^2 + dx^3 \qquad (1.10)$$

We can draw this curve through any four points, say the first four
at $x = 0, 100, 200$ and $300$ m. This gives a good approximation for
the first two intervals. Then we consider fitting the next four points
(starting at $x = 100$ m), with new values for the constants $a, b, c$ and

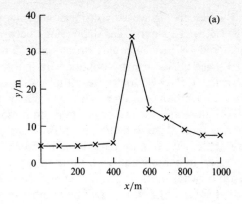

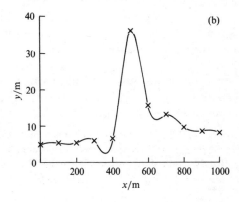

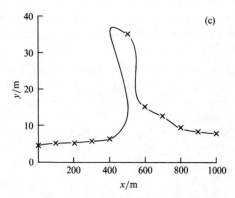

*Figure 1.6*. Approximating a function defined at a limited number of points: (a) by straight lines ('piecewise linear'); (b) by a cubic polynomial ('spline'); (c) by a computational scheme.

*d*, and so on. By these means we can draw a rather smooth curve, as shown in Fig. 1.6(b). This curve has depressions between 300 and 400 m, and 600 and 700 m, which may or may not be real.

But even stranger curves can be obtained by 'curve fitting'. A computer program (GINO F) for curve fitting produced for these points a curve with overhanging edges, which is unlikely (though not impossible). This illustrates the important point that more building blocks do not necessarily lead to better approximations to functions if the data are limited. With more data, building blocks have to be used more sensibly, in conjunction with other information such as one's experience of what hillsides can and cannot look like.

Functions that consist of a sum of a finite number of terms, each of which is a constant times $x^n$, where *n* is a positive integer, are called *polynomials*. The polynomial in equation (1.10) is a *cubic* polynomial. Special polynomials have been worked out with particular constants, such as the so-called Legendre polynomials for the shape of the Earth.

But there are many more functions which can be represented only by an *infinite* sum of terms like $x^n$. These are called *power series*. A few standard series are listed below. Note that the terms often contain factorials ($3! = 3 \times 2 \times 1$, $n! = n \times (n - 1) \times (n - 2) \times \cdots \times 3 \times 2 \times 1$). The trigonometric functions (which we shall meet in later chapters) can be expressed as

$$\sin x = x - \frac{x^3}{3!} + \frac{x^5}{5!} - \frac{x^7}{7!} + \cdots, \quad -\infty < x < +\infty \quad (1.11)$$

$$\cos x = 1 - \frac{x^2}{2!} + \frac{x^4}{4!} - \frac{x^6}{6!} + \cdots, \quad -\infty < x < +\infty \quad (1.12)$$

Try evaluating these series on a calculator. You should find that for $x = \pi/2$, the error in $\sin x$ is less than 0.5% using only three terms. (Note that in these expressions *x* is in *radians*; to convert to degrees multiply *x* by $180/\pi = 57.3$.)

Many functions can be represented as power series in *x*, which we shall call $S(x)$, only over a limited range of values of *x*. It is convenient to write $f(x) = S(x)$ in this interval. For example, if $f(x) = 1/(1 - x)$, then

$$S(x) = 1 + x + x^2 + x^3 + \cdots \quad (1.13)$$

This expansion is valid only for $|x| < 1$ (where $|x|$ is the magnitude, or *modulus*, of *x*). As $x \to -1$, $f(x) \to \frac{1}{2}$, but $S(x)$ oscillates between

$-1$ and $0$ as more terms are taken in the series.* (Try out a few values on your calculator.)

We have seen examples of some building-block functions and how these can be expressed in terms of other, even simpler functions. In a later chapter (Chapter 9), on oscillations, we shall see how sine and cosine functions are used as building blocks to represent waves and vibrations. They can be used in many other applications. In the next section we look at the peculiarities of some rather simple functions.

### Peculiarities of functions and relations: inverse, implicit, impossible and many-valued

The parabola $y = x^2$ is one of the simplest building-block functions we have been considering (see Fig. 1.7(a)). It is also a function of interest in its own right because it represents the trajectory of objects propelled upwards against a gravitational field, if there is no air resistance (e.g. astronauts loping around on the Moon). It is a useful approximation even if there is air resistance, as there is on Earth. By turning the diagram upside down, and defining the ground as $y = 1$ and the position of release of an object as $x = 1$ (with measurements in metres), the equation can be rewritten as $y' = 2x' - x'^2$, where $y' = 1 - y$ and $x' = 1 - x$, the object's angle of departure is $63.4°$ and its initial speed is $5\,\mathrm{m\,s^{-1}}$.

So far we have looked at this equation as giving a value of $y$ for each value of $x$. But if we now ask, 'What is the value of $x$ for a given positive value of $y$?' (or, 'How far along the object's trajectory does it reach a certain height?'), we see from Fig. 1.7(a) that there are *two* answers:

$$x = \sqrt{y} \quad \text{and} \quad x = -\sqrt{y} \qquad (1.14a)$$

This is an example where there is not a one-to-one correspondence (or mapping) between values of $x$ and $y$.

If we follow the rule that the *independent* variable (i.e. the variable we specify) is $x$, and the *dependent* variable (i.e. the one we then calculate) is $y$, then in general the answer to the above question is found by reflecting the curve $y(x)$ in the line $y = x$ (defined in Fig. 1.3) to give the curve

$$y = \pm\sqrt{x} \qquad (1.14b)$$

---

* The arrow here means 'gets closer and closer to'; it does not indicate a functional dependence, as did the arrows $\mapsto$ on p. 7.

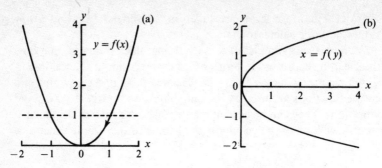

*Figure 1.7.* Parabola and inverse. Diagram (a) shows a parabola as an example of a function $y = f(x)$. By inverting the diagram, and taking the dashed line $y = 1$ as ground level, we obtain the trajectory of a projectile. Diagram (b) shows the inverse relation $x = f(y)$ of the parabola, obtained by reflection in the line $y = x$.

(see Fig. 1.7(b)). This is an example of an *inverse relation*: it is the inverse of the original one in equation (1.8). (Notice that we are dealing with relations here, but not with functions, since for a given value of $x$ the inverse relation gives more than one value of $y$. Also, the inverse relation of $y = \pm\sqrt{x}$ is $y = x^2$. This can be seen either by reflecting $y = x$ again, or by noticing that $(\pm\sqrt{x})^2 = x$.)

### *The parabola $y = \pm\sqrt{x}$ and the 'drunkard's walk'*

The parabolic *curve* defined by equation (1.14b) is widely used because it describes diffusion. For example, if some pollutant is released into a stream at the point $x = 0$, $y = 0$, and the current is flowing from left to right at $1\,\mathrm{m\,s}^{-1}$, the width $W$ of the polluted 'plume' is described by $W \approx \sqrt{x}$ (where the sign $\approx$ means 'approximately equal to'). This would be the width you would measure on a time-exposure photograph, not the width at an instant of time).

In fact, many processes like diffusion can be 'modelled' by a 'drunkard's walk' – a lurch to the left ($s > 0$), or a lurch to the right ($s < 0$), for each step forward $r$ ($>0$) (see Fig. 1.8). This is a model of small, unconnected random movements. At, say, the $n$th step, the lurch to the left or right is denoted by $s_n$. Each value of $s_n$ is independent of every other value: each step taken has nothing to do with where the drunkard has come from or where he is going. After

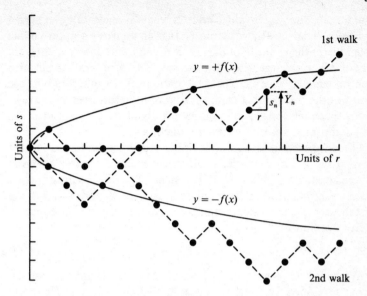

*Figure 1.8.* Examples of 'drunkard's walks' across a street. $Y_n$ is the distance from the $x$ axis after $n$ steps. The curves $y = \pm f(x)$ are found from the average of $Y_n^2$ over many drunkard's walks.

$n$ steps the net movement of the drunkard is $Y_n = s_1 + \cdots + s_n$, so the average value of $Y_n$ is zero. But the square of $Y_n$ is

$$Y_n^2 = s_1^2 + s_2^2 + \cdots + s_n^2 + 2(s_1 s_2 + s_2 s_3 + \cdots$$

$$+ \text{ all other products of different displacements}) \quad (1.15)$$

Now we consider the value of this sum over many steps and over many different drunken walks. Since the drunkard is so drunk that his lurch $s_1$ has nothing to do with subsequent lurches $s_2, s_3, \ldots$, on average the products $s_1 s_2, s_{10} s_{13}, \ldots$ are zero. We assume each lurch has the same magnitude to the left or right (of $\pm s$). Then the average of $Y_n^2$ is $ns^2$. But in $n$ lurches, the drunkard has moved forward a distance $x = nr$ steps, and so $n = x/r$. Therefore the average of $Y_n^2$ at $x$, which we call $y^2(x)$, can be expressed in terms of the distance $x$ and the lengths of the drunkard's forward and sideways steps ($s$ and $r$) as

$$y^2 = x(s^2/r) \quad (1.16)$$

The value of $y$ tells us how much of the street he is wandering over. Here $y$ is defined for a finite number of steps. This drunkard's walk is similar to the way in which eddy currents spread pollution across a stream. That is why equation (1.16) for $y$ gives an approximate estimate for the width $W$ of the plume of pollutant downstream from where it is discharged into the stream. The drunkard's walk also approximates the way molecules bump into each other and diffuse heat or other molecules through a solid. But in this case there are so many tiny 'collisions', and $r$ and $s$ are so small, that we consider $y(x)$ as an approximation to a *smooth* function.

Often we need to look at how diffusion proceeds with *time*, rather than *distance*. For example, a substance (such as a trace element in a semiconductor, or water in the stem of a plant) can diffuse a distance $D$ in a time $t$ by many collisions of period $r$ over small distances $s$. By the same argument we used to arrive at equation (1.16), we find that

$$D^2 = tK, \quad \text{where } K = s^2/r \qquad (1.17)$$

We have seen here a simple example of a mathematical function being developed via a mathematical 'model' of a process. There will be many other examples later. Also, we have seen the same function emerging from two mathematical models of two different physical situations.

### *Implicit form*

The description of the diffusion process has shown us an equation relating the dependent variable $y$ to the independent variable $x$ which is not in the form $y = f(x)$. Instead, a function of $y$, say $g(y)$, is equal to a function of $x$:

$$g(y) = f(x) \qquad (1.18)$$

In our example of equation (1.16), $g(y)$ is $y^2$ and $f(x)$ is $x(s^2/r)$.

If we slice up cones in different ways, as shown in Fig. 1.9, we obtain 'conic sections' – familiar and important curves which are usually represented in the form of equation (1.18). These curves are circles, ellipses and hyperbolas, mathematical 'cousins' of the parabolas (also conic sections) we have just been discussing. Examples of each of these types of curve may be obtained from the general equation

$$y^2 = 1 - \mu x^2 \qquad (1.19)$$

where $\mu$ is a constant, or *parameter*; $\mu$ may take any value between $-\infty$ and $+\infty$, but is constant for any particular curve of $y$ against

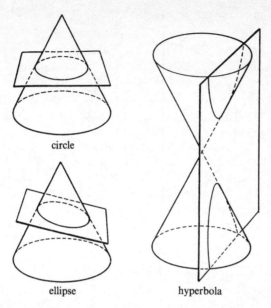

circle

ellipse                              hyperbola

*Figure 1.9.* Circles, ellipses and hyperbolas are slices through
cones.

*x*. All the curves represented by equation (1.19) pass through the
points $(0, \pm 1)$. The range of values of the independent variable $x$ for
which the dependent variable $y$ can be defined depends on the value
of $\mu$ (see Fig. 1.10). For $\mu = 1$ this equation represents a circle of
radius 1, so $y$ can be calculated only for $-1 \leqslant x \leqslant 1$. (This can be
expressed more succinctly, in terms of the modulus of $x$, as $|x| \leqslant 1$.)
Can you think of four good reasons why similar shapes are so
common in everyday products? Chapter 4 will tell you!

For $\mu > 0$, equation (1.19) represents ellipses, which can be
classified as flat ellipses for $0 < \mu < 1$ and elongated ellipses for
$1 < \mu < \infty$. These curves correspond to different slices across the
cone, or slices across a circular pipe, or they can represent, say, the
motion of a planet round the Sun, or the value of a variable (such
as the approximate population of prey in a simple biological predator-
prey system) – see Chapters 5 and 7. The ranges of values of $y$ and
$x$ are $|y| < 1$ and $|x| < 1/\mu$.

What do you think happens as $\mu$ gets smaller and smaller ($\mu \to 0$)
or larger and larger ($\mu \to \infty$)? The algebraic answer comes from
equation (1.19). As $\mu \to 0$, $y^2 \to 1$ and so $y \to \pm 1$, and therefore we
have two straight lines (see Fig. 1.11). Geometrically this limit is
equivalent to flattening out the ellipse to an infinite extent, since we

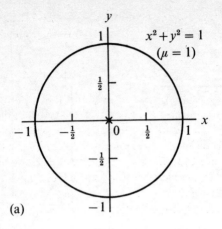

(a)

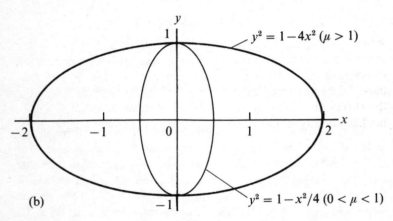

(b)

*Figure 1.10.* Circles and ellipses have equations of the form
$$g(y) = f(x).$$

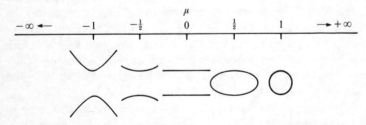

*Figure 1.11.* As $\mu$ in the equation $y^2 = 1 - \mu x^2$ is varied,
different curves appear. What will happen as $\mu$ approaches
$\pm\infty$?

have shown that these ellipses extend a distance $\pm 1/\mu$ along the $x$ axis, as $\mu \to 0$. You should be able to work out the limit as $\mu \to \infty$.

### Limits

Philosophers always want to take an argument to its final logical conclusion, no matter how silly the answer, because that way they can get a better understanding of the *assumptions* they make. Users of mathematics are the same: they want to know where their model or their equation leads to in the most extreme circumstances – here, with our ellipses, this has meant letting parameters tend to zero or infinity. As in philosophy, it is one of the best ways of exploring assumptions or methods of calculation.

### Closed and open curves

When $\mu < 0$, equation (1.19) generates hyperbolas (see Fig. 1.12), so the forms of the curves have suddenly changed from *closed curves* to *open curves*. These curves do not close on themselves, and (for finite values of $\mu$) they extend to $\infty$. That is, values of $x$ and $y$ on the curves are 'unbounded' – a better term than 'infinity', which mathematicians find too imprecise for many kinds of analysis.

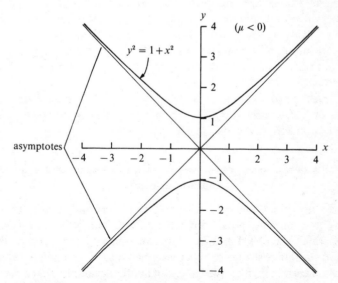

*Figure 1.12.* Hyperbolas also have equations of the form $g(y) = f(x)$.

## Asymptotic forms

An important feature of hyperbolas, as for many curves and for many functions, is their form at infinity. Looking at equation (1.19), we see that when $y$ and $x$ are large (say greater than 10) and, say, $\mu = -1$, we can approximate it by the simpler equation

$$y^2 \approx x^2 \qquad (1.20a)$$

Therefore, at any value of $x$, there are two values of $y$:

$$y = x \quad \text{and} \quad y = -x \qquad (1.20b)$$

As we saw in Fig. 1.4, the graphs of these two equations are straight lines. These lines are called the *asymptotes* of the hyperbola, because the actual solution for $y$ (or curve), defined by equation (1.19), tends *asymptotically* to the approximate solution (1.20b). Notice how rapidly the curve approaches its asymptotic form.

So, we have found that looking at the *limiting* or asymptotic form of a solution or a curve gives a useful approximation to the curve over a wide range of values. The commonly held view that mathematics consists largely of accurate calculations with numbers is, of course, a great misapprehension. Most of the time mathematicians are making approximations, doing approximate calculations, and looking for the asymptotic or most peculiar aspects of a curve or a function. This is because it is easier to think about and understand simpler forms, equations or algebraic expressions or curves. (It is not usually a case of being lazy; in fact, thinking about the best simplifications and explanations usually takes much longer than computing the answer!) We shall see plenty of examples in later chapters.

Curves approximating to hyperbolas can be seen in the shapes of some shell structures and flexible structures, like cooling towers and roofs of airport buildings.

### Tangents, speed traps, cricket balls, roller coasters and a touch of calculus

In our first scene, Lee Brown wanted to understand curves of flow rate $F$ at different depths $x$ down the oil well by plotting the *change* in $F$, which she called $q$, for each metre along the pipe. One of the important things she noted was that $q$ had maximum and minimum values at certain depths. We have seen in other examples that a curve can be closely approximated by a series of straight lines drawn between points on the curve, e.g. $y(x)$, and that the greater the number of points (or, which amounts to the same thing, the smaller the distance between them), the closer the straight lines come to the curve.

We hope you are now equipped to follow, in an informal way, the great leap into calculus made by Isaac Newton and Gottfried Leibniz in the seventeenth century. Calculus remains central to the application of mathematics in the twentieth century. But with computers and the modern way of always looking very closely at different ways of measuring and analysing, new approaches have been used.

### Tangents to curves

First we need to be reminded that the *tangent* to a curve at a point, say $(1, 1)$ on the curve $y = x^2$ in Fig. 1.13(a), is the straight line through the point which just *touches* the curve at that point. Mathematically, 'touching' means that the tangent line has the same *slope* as the curve. We defined the slope of a straight line in equation (1.7). But what is the slope of a curve?

Calculus provides the rules for calculating this tangent from what we know about the function $y = f(x)$ near the point $x = x_0$, $y = y_0$. Since drawing a tangent seems rather undefined, a first suggestion for a definite procedure might be to take a ruler and draw a line between two points on the curve near each other: $(x_0, y_0)$ and

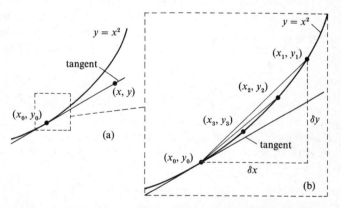

*Figure 1.13.* Defining the slope of a curve and the derivative as limiting processes. (a) The tangent to the parabola $y = x^2$ at $(x_0, y_0)$. (b) Enlargement of the region near $(x_0, y_0)$ showing lines drawn between $(x_0, y_0)$ and the points $(x_1, y_1)$, $(x_2, y_2)$ and $(x_3, y_3)$. For points closer to $(x_0, y_0)$, the lines are closer to the tangent.

another point $(x_1, y_1)$. If these two points are close, the line 'looks' very similar to the tangent. In fact, the slope of the line, $m_1$, is

$$m_1 = \frac{y_1 - y_0}{x_1 - x_0}$$

or, in the function notation,

$$m_1 = \frac{f(x_1) - f(x_0)}{x_1 - x_0}$$

For example, if $x_1 = 3/2$, $y_1 = 9/4$, so $m_1 = 2.5$. (This is close to what Lee Brown did: she calculated the small change in $F$ for a small increase in $x$.) We denote by $\delta y$ this small change, $y_1 - y_0$, in $y$, and by $\delta x$ the change $x_1 - x_0$ in $x$. (These small quantities are usually called *infinitesimals*.)

Now, what would happen if we had drawn a line to a different point, $(x_2, y_2)$, close to $(x_0, y_0)$, as in Fig. 1.13(b)? Would the slope of the estimated tangent $m_2 = (y_2 - y_0)/(x_2 - x_0)$ be the same? Suppose that $x_2 = 1.1$; then $y_2 = 1.21$, and $m_2 = 2.1$. So here, and indeed for most functions, the answer is 'Almost'. As $\delta x$ becomes smaller and smaller, so does $\delta y$, and thus the ratio $\delta y/\delta x$ tends to a definite limit, the slope or *gradient* of the graph at $(x_0, y_0)$, denoted by $dy/dx$. We call this the *differential* (or *derivative*) of $y = f(x)$ with respect to $x$. We can express this result in symbols, as an equation:

$$\frac{dy}{dx} \text{ (at } x = x_0) = \lim_{\delta x \to 0} \frac{\delta y}{\delta x} \text{ (at } x = x_0) \qquad (1.21)$$

Another way of writing this is

$$\frac{dy}{dx} \text{ (at } x = x_0) = \lim_{x \to x_0} \frac{f(x) - f(x_0)}{x - x_0}$$

One or two applications demonstrate how this definition of $dy/dx$ is used. The tangent that touches a straight line $y = ax + b$ is the same as the straight line itself and has the same slope, given by equation (1.7), so

$$\frac{dy}{dx} = a \qquad (1.22)$$

But for the parabola $y = x^2$ in Fig. 1.13 the tangent is quite distinct from the curve. To calculate $dy/dx$, we first consider the small finite changes in $y$ when $x$ changes by $\delta x$. Then

$$\frac{\delta y}{\delta x} = \frac{(x_0 + \delta x)^2 - x_0^2}{\delta x} = 2x_0 + \delta x \qquad (1.23)$$

So, taking the limit as $\delta x \to 0$, this equation becomes, at a general point $x$,

$$\frac{dy}{dx} = \lim_{\delta x \to 0} (2x + \delta x) = 2x \tag{1.24}$$

Thus the slope of the parabola changes sign with $x$ and increases as $x$ increases. At $x = 1$, $dy/dx = 2$ (which is close to the value we obtained for the slope $m_2$ of the estimated tangent lines).

### Velocity

We can use calculus to define and calculate the velocity of an object moving at variable speed. (Have you heard the one about the policeman who stopped the motorist and said, 'You were doing 40 miles an hour'? The motorist replied, 'But Officer, I could not have been, because I've not been driving for an hour. I've only driven 2 miles in 3 minutes.') For an object moving in a straight line, which has reached a position $x$ at a time $t$, the velocity is defined as the change in position, $\delta x$, over an infinitesimal interval of time, $\delta t$:

$$v = \lim_{\delta x \to 0} \frac{\delta x}{\delta t} = \frac{dx}{dt} \tag{1.25}$$

Since $v$ can be defined for a range of values of $t$, it can now be regarded as a function of time: $v(t)$. (So the policeman's answer should have been, 'But I [or his instruments] have differentiated your distance with respect to time and found your speed to be 40 miles per hour at the time in question, so you'd better show me your driving licence and answer a few questions!')

Sometimes we know the slope of a curve, but want to work out what the curve is. Perhaps we know approximately how a cricket ball accelerates downwards, but want to find its speed and where it goes.

Acceleration, $a$, is defined as the small change in velocity over an infinitesimal interval of time, so it is calculated by differentiating $v(t)$:

$$a = \frac{dv}{dt} \tag{1.26}$$

A particle propelled through the air near the Earth's surface will have (if we neglect air resistance) an upward vertical acceleration of $-g$, or approximately $-10 \, \text{m s}^{-2}$. From equation (1.26), the vertical velocity $v$ of this particle must be steadily decreasing, i.e.

$$\frac{dv}{dt} = -g = -10 \tag{1.27}$$

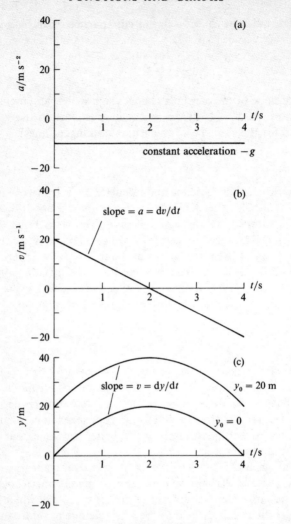

*Figure 1.14.* The derivative of a function, or curve, is another function or curve: (a) shows the acceleration, (b) the velocity and (c) the height of a projectile; (b) is the derivative of (c), and (a) is the derivative of (b).

(see Fig. 1.14(a)). Since the only curve which has a fixed gradient is a straight line, the graph of $v(t)$ against $t$ must be a straight line. The slope of the graph is $-10$. The equation of the straight line is

$$v(t) = b - 10t \qquad (1.28)$$

but we don't know $b$. However, if the particle is initially (i.e. at $t = 0$) projected upwards with a velocity $20 \, \mathrm{m \, s^{-1}}$, then by putting $t = 0$ in equation (1.28), we find that $b = 20$. Therefore, the expression for the function $v(t)$ is

$$v = 20 - 10t \qquad (1.29)$$

(see Fig. 1.14(b)). We say this is the *solution*, or the *integral*, of equation (1.27). (We return to integrals in Chapter 4.)

Equation (1.29) means that at time $t = 20/10 = 2 \, \mathrm{s}$, the velocity $v = 0$ and the particle stops moving and starts to drop. We can now calculate the height of the particle $y(t)$, because $v$ is related to $y(t)$ by

$$v = \frac{\mathrm{d}y}{\mathrm{d}t} = 20 - 10t \qquad (1.30)$$

The solution to equation (1.30) is not simple because $\mathrm{d}y/\mathrm{d}t$ varies with $t$. What we do is *guess* a solution for $y(t)$ so that when we differentiate it $\mathrm{d}y/\mathrm{d}t$ satisfies equation (1.30):

$$y = 20t - 5t^2 + y_0 \qquad (1.31)$$

Note that $y_0$ can be any constant because its value does not affect $\mathrm{d}y/\mathrm{d}t$. Different values of $y_0$ are shown in Fig. 1.14(c). If the particle is released at the ground, $y = 0$ when $t = 0$, so $y_0 = 0$.

We notice from this curve that at the *top* of the trajectory, where $t = 2$, the velocity $v = 0$ and the slope $\mathrm{d}y/\mathrm{d}t = 0$. (This is like the point that Lee noticed in her analysis of the oil well curve, $q(x)$.) The general point is that for any curve, the places where $\mathrm{d}y/\mathrm{d}x$ (or $\mathrm{d}y/\mathrm{d}t$) are zero are usually places where the curve has a local *maximum* or *minimum*. (Can you think of a curve where $\mathrm{d}y/\mathrm{d}x = 0$ at a point which is not a maximum or minimum? How about $y = x^3$ at $x = 0$?) We'll illustrate this further with a familiar example of ups and downs – over hills, say, or a roller coaster – as shown in Fig. 1.15.

The mathematical meaning of the term 'maximum' is that the $y$ value of the curve or function is greater than it is over a certain distance either side of the maximum. It might be the top of a small hill B, near a much larger one, as shown in Fig. 1.15(a). At a minimum, such as C, the function has the smallest value over a certain distance. At both places the slope is zero, which means that the first (small) step you take away from the top or bottom of the hill is virtually horizontal (i.e. $\mathrm{d}y/\mathrm{d}x \approx 0$).

But what is the mathematical difference between the bottom of a valley and the top of a hill, since at each place $\mathrm{d}y/\mathrm{d}x = 0$? Well, you

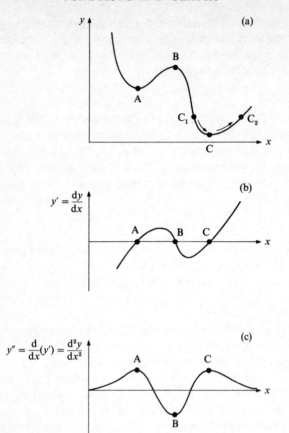

*Figure 1.15.* Maximum and minima of a function $y$, shown
in diagram (a), and the behaviour of the derivatives (b) $y'$
and (c) $y''$ at the same positions.

go *down* into a valley and *up* the other side, so as you approach
the point C from the point $C_1$ in Fig. 1.15(a), $dy/dx$ (or $y'$) is
negative, but changes to positive as you start climbing away from C
towards $C_2$. The derivative $dy/dx$ is plotted as another function in
Fig. 1.15(b). Things are different as you pass over the little hilltop at
the maximum B, where $dy/dx$ is first positive and then negative.

With our new-found knowledge of calculus, we will naturally seek
to understand Fig. 1.15(b) better by plotting a graph of the slope of
$y'$, i.e. $dy'/dx$. (This is usually denoted by $d^2y/dx^2$, or by $y''$, because

now $y(x)$ has been differentiated twice.) Notice that at the point B of the maximum (in $y$), where $y'$ is decreasing from positive to negative, $d^2y/dx^2$ is negative ($d^2y/dx^2 < 0$). At the two points of minima in $y$ (A and C), $y'$ is increasing and so $d^2y/dx^2 > 0$.

In physics, chemistry, engineering and biology, many of the 'laws' are expressed in terms of these differential coefficients, or derivatives – usually up to 'second order' (i.e. there are terms like $dy/dx$ or $d^2y/dx^2$) but sometimes to higher order. The reason is that this is the most *general* and concise way of expressing the law.

For example, if we consider the accelerating particle in Fig. 1.14, the law of acceleration (Galileo's) that governs the particle is that $a = -g$. Since $a = dv/dt$ and $v = dy/dt$, the 'law' for the variation of the height of the particle with time, $y(t)$, can be expressed in the most general way as

$$a = d^2y/dt^2 = -g \tag{1.32}$$

or, in words, 'the second derivative of the height of the particle is equal to minus $g$'. This is more general than stating an equation for $y(t)$ in the form of a parabolic curve, because there are many possible parabolas for the different starting positions and starting velocities of the particle, as we can see for two of them in Fig. 1.14(c).

The forces you feel on your stomach as you ride on a roller coaster illustrate why Newton's famous second law is often expressed in terms of the second derivative. The force $F$ on an object (your stomach) is given by $F = ma$, where $m$ is the object's mass and $a$ is the acceleration. So the vertical force is $F = m\,d^2y/dt^2$. For a roller coaster, the curve of $y(t)$ has approximately the same shape as $y(x)$, shown in Fig. 1.15(a). So at the points A and C, where $d^2y/dt^2$ is positive, the acceleration of your body is upwards; for your stomach to accelerate upwards it requires an upward force from the body. This is what you feel, as it seems that your stomach is forced into your legs. At the point B, where $d^2y/dt^2$ acts downwards, the downward force on your stomach seems to be exerted by your throat; your stomach almost feels as if it is in your mouth.

We shall see other examples of general laws in Chapters 7 and 9.

To return to Scene 1 at the start of this chapter, perhaps it is clearer now why Izzy and Lee found it easier to interpret their curve of $F$ against $x$ by approximately differentiating $F$. They were estimating the function $q(x)$ defined by $q(x) = dF/dx$. When you have done this a few times, you will be able to differentiate curves with your eyes – you won't need to use your head at all!

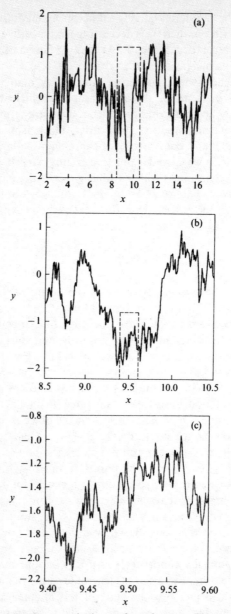

*Figure 1.16.* A very wiggly, or *fractal*, curve. The usual rules of differentiation cannot be applied because the curve remains wiggly on smaller and smaller scales. On going from (a) to (b) to (c), the *x* scale is stretched by a factor of roughly 10 each time, but the shape of the curve remains similar.

## Very wiggly curves

So far in this chapter we have been looking at rather smooth curves or functions. However, in reality many functions are rather like the one plotted in Fig. 1.16: they have many maxima and minima close together, and if you look closely at any particular part of the curve, you will find even more maxima and minima. This is what you would see if you tried to define a surface of a piece of metal, or a coastline, or the outline of a cloud, or if you measured the velocity of water in a river. In this case, the slope $dy/dx$ at a point cannot be defined in terms of limits: we could not draw the kind of diagram we did in Fig. 1.13.

However, we often need to know the shape and slope of a surface (of a hill or a piece of metal). Since we will get nowhere looking at it in microscopic detail, we consider instead an approximate smooth curve based on smoothing the shape over a certain *scale*. Just like drawing a tangent to a curve, this can be done either by 'eye' or, as we shall see in Chapter 4, in a systematic way.

There are many fascinating problems in science and technology, as well as in mathematics, where these kinds of 'wiggly' curve, or *fractals*, appear. It is beyond the scope of this chapter to discuss them (but see Chapter 6).

### Functions meeting, evolving and having a catastrophe

In Scene 2 at the start of this chapter, Hal drew two graphs on the same diagram to explain that, once the curves of costs and sales cut each other, the business would be in profit.

One of the main uses of functions is to show relations between variables – in this case outgoings $O$ and sales $S$, each of which are *functions* of one other variable – time. So Hal plotted $O$ and $S$ as functions of time, $O(t)$ and $S(t)$, and wanted to show that it would be possible for the sales curve $S(t)$ to cross the outgoings curve $O(t)$. If the curves crossed, they would be in profit.

The mathematical question is, 'Will there be a time $t$ when $O(t) = S(t)$?' In any particular case, the answer to this question must depend on the form of the functions, which we can explore again by looking at their graphs.

So let us look briefly at the question of when curves do and do not cut each other, and at what kinds of application there are for this property of functions. The simplest case is an intersection of two straight lines, as shown in Fig. 1.17. The straight line passing

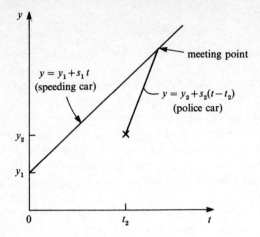

*Figure 1.17.* The intersection of two straight-line functions:
a car chase.

through $y$, at time $t = 0$, has the equation $y = y_1 + s_1 t$. Let this represent a car travelling at an excessive speed, $s_1$. It is spotted at time $t_2$ at position $y_2$ by a police car. If the police car travels at speed $s_2$, when does it catch the speeding car? The straight line representing the police car has the equation $y = y_2 + s_2(t - t_2)$. When the two curves intersect, i.e. the cars meet at the same place and the same time, the value of $y$ and the value of $t$ will be the same for the two curves. Therefore

$$y = y_1 + s_1 t = y_2 + s_2(t - t_2) \qquad (1.33)$$

We can now find this value of $t$ in terms of our original information $y_1$, $s_1$, $y_2$, $s_2$ and $t_2$. The answer (obtained by rearranging the terms in equation (1.33)) is

$$t = \frac{y_2 - y_1 - s_2 t_2}{s_1 - s_2} \qquad (1.34)$$

Having the answer in symbols (or 'algebraic form') means that it is easy to see at a glance what the solution means in different cases. For example, we can see that the interception time $t$ increases when $s_2 - s_1$ decreases: that if the police car is only just faster than the speeding car, the chase will be a long one!

This simple calculation of the intersection of straight lines is the basis of more complex calculations – of, for example, the formation of flood surges along rivers caused by tides or heavy rainfall.

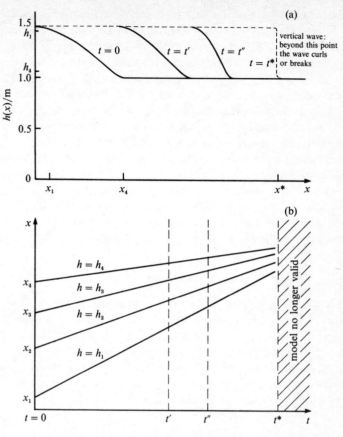

*Figure 1.18*. A rising tide leads to catastrophe. (a) Change in the water surface with time. It eventually becomes vertical: the wave breaks and the model breaks down. (b) Graph showing how points on the surface, the curve $h(x)$, move as the wave advances.

Take the case of a tide rising in a river. Suppose that, without tidal disturbance, the depth of the river is a uniform 1 metre, and is increased to 1.5 metres by the tide. This is illustrated in Fig. 1.18(a). Here $x$ is the distance up-river. The velocity of the flowing water depends on depth. The theory of disturbances and waves on water shows that the greater the depth of the water, the greater will be the velocity; thus deeper water catches up with shallower water. In Fig. 1.18(a) the profile of the water surface becomes steeper at later times $t = t'$ and $t = t''$.

In Fig. 1.18(b) there are lines to show where the river is at depths $h_1$, $h_2$, $h_3$ and $h_4$, at later times. On this $(t, x)$ diagram the gradient of the line for $h = h_1$ is the speed of the water at that depth. Water at greater depths travels faster, and the greater the depth, the steeper the line. Hence the straight lines come closer together, corresponding to the water surface becoming steeper. Eventually these lines may meet at a point, indicating that the surface has become vertical. Of course, our model no longer applies at the later stage when the wave breaks.

The extreme values are shown on the diagrams as $x^*$ and $t^*$. In a typical situation $t^*$ might be 3 hours, and $x^*$ might be 30 kilometres. This is an example of a *catastrophe*, where functions may suddenly change or have derivatives which are infinite. By looking at the mathematical aspects of different kinds of catastrophe, mathematicians now have a better understanding of how catastrophes develop in biology, engineering and many other disciplines.

## Summary

We have seen in this chapter that functions are the key to mathematical modelling of everyday activities. Many scientific studies and engineering calculations also depend on the use of functions. We can usually assume that functions of variables are well-behaved and vary smoothly, but we must always watch out in case functions have more than one value, or change 'catastrophically', or are very 'wiggly' – even under a mathematical microscope!

## Further reading

I. Stewart, *Concepts of Modern Mathematics* (Penguin, 1975).
I. Stewart, *The Problems of Mathematics* (Oxford University Press, 1987).
A. Woodcock and M. Davis, *Catastrophe Theory* (Penguin, 1980).

# 2

## SOLVING LINEAR EQUATIONS

### NICK HIGHAM

**A problem . . .**

A scientist friend approaches you with the following problem: From
some experimental data, concentrations of a chemical measured at
one-second intervals (see Table 2.1), can you estimate the concentrations at various intermediate times? You plot the points on a
graph (Fig. 2.1) and explain that one possible approach is to sketch
a curve which passes through the data points. The required estimates
can then be read from the curve. However, since neither of you is
very good at drawing curves by hand, you decide to tackle the
problem mathematically.

To begin, you consider just a few points, hoping that it will
become clear how to proceed with all ten. If there were just two
points, you could take the curve to be a straight line. This seems too
easy, so you consider three points. A line which is to pass through
three arbitrary points must have some curvature, and the simplest
such curve is a parabola. You recall that the general formula for a
parabola is

$$y = at^2 + bt + c \qquad (2.1)$$

where $a$, $b$ and $c$ are constants, $t$ is the independent variable and $y$
the dependent variable. To find $a$, $b$ and $c$ you'll have to express,
mathematically, the requirement that the curve passes through the
first three data points; that is,

$$
\begin{aligned}
y &= 3 \quad \text{for } t = 1 \\
y &= 5 \quad \text{for } t = 2 \qquad (2.2)\\
y &= 1 \quad \text{for } t = 3
\end{aligned}
$$

*Table 2.1.* Experimental data.

| Time (seconds) | Concentration (moles) |
| --- | --- |
| 1 | 3.0 |
| 2 | 5.0 |
| 3 | 1.0 |
| 4 | 2.5 |
| 5 | 2.0 |
| 6 | 1.5 |
| 7 | 1.0 |
| 8 | 0.8 |
| 9 | 0.7 |
| 10 | 0.6 |

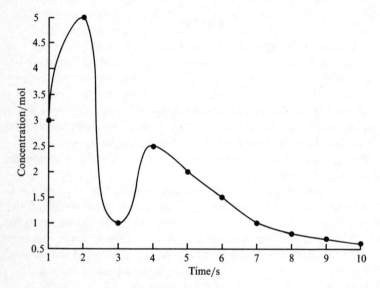

*Figure 2.1.* Plot of the data given in Table 2.1, with a curve drawn through the data points.

Writing out these equations in detail you arrive at

$$c + b + a = 3 \tag{2.3}$$

$$c + 2b + 4a = 5 \tag{2.4}$$

$$c + 3b + 9a = 1 \tag{2.5}$$

Recognizing that this is a set of three simultaneous equations in three unknowns, and recalling the method of 'elimination of variables', you carry out the following sequence of operations. First, you subtract equation (2.3) from equations (2.4) and (2.5) to obtain the new equations

$$b + 3a = 2 \tag{2.6}$$

$$2b + 8a = -2 \tag{2.7}$$

in which $c$ has vanished. Next, you subtract twice equation (2.6) from equation (2.7) to get

$$2a = -6 \tag{2.8}$$

Equation (2.8) shows that $a = -3$. Substituting this value for $a$ into equation (2.6) gives $b - 9 = 2$, or $b = 11$, and then from equation (2.3) $c + 11 - 3 = 3$, or $c = -5$. To confirm that no computational slips have been made you check that these values of $a$, $b$ and $c$ satisfy equations (2.4) and (2.5).

You conclude that the quadratic

$$y = -3t^2 + 11t - 5 \tag{2.9}$$

passes through the data points, and you tell your friend to evaluate $y$ from equation (2.9) for $t = 1.5, 2.5$, and so on to get estimates of the required concentrations.

Impressed by your mathematical dexterity, your friend reminds you that the original problem has 10 data points. Realizing that this problem would require the solution of 10 equations in 10 unknowns – a formidable task to undertake by hand – you make your excuses and suggest that your friend finds a computer of the silicon kind!

## Vectors and matrices

The elimination process illustrated above is quite straightforward to apply when the number of equations and unknowns, $n$, is just 2 or 3. But solving bigger systems of equations by hand is more difficult. As $n$ increases, the amount of algebra (and hence the amount of paper needed) grows rapidly, and in any case it is perhaps not clear what the general method is for $n > 3$. Some of these difficulties are eased by using vector and matrix notation, which we now introduce. We shall use this notation when we come to describe the method of *Gaussian elimination* for solving $n$ equations in $n$ unknowns. (If you are familiar with the language of vectors and matrices, you can skip to p. 39.)

A *vector* is a column of numbers. Some examples are

$$u = \begin{pmatrix} 2 \\ -1 \\ 3 \end{pmatrix}, \quad v = \begin{pmatrix} 5 \\ 0 \\ -4 \\ 1 \end{pmatrix}, \quad w = \begin{pmatrix} 3 \\ -7 \end{pmatrix} \quad (2.10)$$

Strictly these are called *column vectors*; there are also *row vectors* in which the numbers are arranged in a row, as we'll see shortly. (Note that we use a bold typeface to distinguish vectors from scalars.) The vector $u$ can be thought of as representing the point with coordinates $(2, -1, 3)$ in three-dimensional space.

Vectors provide a useful way of collecting together and naming a string of numbers, but their power becomes apparent only when rules are given for manipulating them. The simplest operation is multiplication of a vector by a scalar, defined as the operation of multiplying each component of the vector by the scalar. For the vectors (2.10),

$$2u = \begin{pmatrix} 4 \\ -2 \\ 6 \end{pmatrix}, \quad -3v = \begin{pmatrix} -15 \\ 0 \\ 12 \\ -3 \end{pmatrix}, \quad 10w = \begin{pmatrix} 30 \\ -70 \end{pmatrix} \quad (2.11)$$

For a point in three-dimensional space with position vector $x$, the vector $2x$ represents the point lying in the same direction as seen from the origin, but at twice the distance.

The next operation to consider is the addition of two vectors. Vector addition is defined only for vectors having the same number of components. It is done in the natural way, component by component, for example:

$$\begin{pmatrix} 3 \\ 2 \\ 9 \end{pmatrix} + \begin{pmatrix} -7 \\ 6 \\ 4 \end{pmatrix} = \begin{pmatrix} -4 \\ 8 \\ 13 \end{pmatrix}$$

$$\begin{pmatrix} 1 \\ -1 \\ 1 \\ 0 \end{pmatrix} + \begin{pmatrix} 3 \\ 2 \\ 0 \\ -1 \end{pmatrix} = \begin{pmatrix} 4 \\ 1 \\ 1 \\ -1 \end{pmatrix} \quad (2.12)$$

You may like to check the consistency of our definitions by comparing $v + v$ and $2v$.

The third operation between vectors that we'll need gives what is called the *inner product* (or *scalar product*, or *dot product*). If

$$p = \begin{pmatrix} -2 \\ 1 \\ 6 \end{pmatrix}, \qquad q = \begin{pmatrix} 4 \\ -1 \\ -2 \end{pmatrix} \qquad (2.13)$$

then the inner product of $p$ and $q$ is

$$p^{\mathrm{T}}q = (-2 \quad 1 \quad 6)\begin{pmatrix} 4 \\ -1 \\ -2 \end{pmatrix} = -8 - 1 - 12 = -21 \qquad (2.14)$$

(In the alternative 'dot product' notation, we would write $p \cdot q = -21$.) The inner product of two vectors, then, is a number, obtained by forming the sum of the products of corresponding components. In (2.14) we used the transpose operator, T, which turns a column vector into a row vector (and vice versa), and we employed the rule

$$(a \quad b \quad c)\begin{pmatrix} x \\ y \\ z \end{pmatrix} = ax + by + cz \qquad (2.15)$$

for multiplying a row vector and a column vector.

Looking at equations (2.3)–(2.5) with an eye to vector notation, we can rewrite them in the form

$$c\begin{pmatrix} 1 \\ 1 \\ 1 \end{pmatrix} + b\begin{pmatrix} 1 \\ 2 \\ 3 \end{pmatrix} + a\begin{pmatrix} 1 \\ 4 \\ 9 \end{pmatrix} = \begin{pmatrix} 3 \\ 5 \\ 1 \end{pmatrix} \qquad (2.16)$$

provided we interpret '$x = y$' as meaning that $x$ agrees with $y$ in each component. Equation (2.16) shows that we have to find a combination of the three vectors on the left that gives the vector on the right. Another interpretation is that we have to travel from the origin $(0, 0, 0)$ to $(3, 5, 1)$, but can move only in the directions specified by the vectors $(1\ 1\ 1)^{\mathrm{T}}$, $(1\ 2\ 3)^{\mathrm{T}}$ and $(1\ 4\ 9)^{\mathrm{T}}$.

Now, the left-hand side of each of equations (2.3)–(2.5) is an inner product between two vectors, so we can rewrite them as

$$(1 \quad 1 \quad 1)\begin{pmatrix} c \\ b \\ a \end{pmatrix} = 3 \qquad (2.17)$$

$$(1 \quad 2 \quad 4)\begin{pmatrix} c \\ b \\ a \end{pmatrix} = 5 \qquad (2.18)$$

$$(1 \quad 3 \quad 9)\begin{pmatrix} c \\ b \\ a \end{pmatrix} = 1 \qquad (2.19)$$

The vector $x = (c\ b\ a)^{\mathrm{T}}$ of unknowns is common to each equation, so we can write (2.17)–(2.19) in a more compact form as

$$\begin{pmatrix} 1 & 1 & 1 \\ 1 & 2 & 4 \\ 1 & 3 & 9 \end{pmatrix}\begin{pmatrix} c \\ b \\ a \end{pmatrix} = \begin{pmatrix} 3 \\ 5 \\ 1 \end{pmatrix} \qquad (2.20)$$

which we define as meaning the same as (2.17)–(2.19).

The array

$$A = \begin{pmatrix} 1 & 1 & 1 \\ 1 & 2 & 4 \\ 1 & 3 & 9 \end{pmatrix} \qquad (2.21)$$

is an example of a *matrix*. In general, a matrix is any rectangular array of numbers, such as

$$\begin{pmatrix} 1 & 2 \\ 3 & 4 \end{pmatrix}, \quad \begin{pmatrix} 5 & 6 \\ 9 & -3 \\ 0 & -1 \end{pmatrix}, \quad \begin{pmatrix} 7 & -3 & 2 & 9 \\ 0 & 11 & -4 & 2 \end{pmatrix} \qquad (2.22)$$

A vector is actually a special type of matrix, one with a single column (or row). Apart from vectors, we are concerned here only with *square matrices* – those that have an equal number of rows and columns.

In equation (2.20) we introduced indirectly the notion of matrix–vector multiplication. The rule for carrying out the multiplication is that if $Ax = b$, then the $i$th component of $b$ is the inner product of the $i$th row of $A$ with the vector $x$. The general formula for $3 \times 3$ matrices is

$$Ax = \begin{pmatrix} a_{11} & a_{12} & a_{13} \\ a_{21} & a_{22} & a_{23} \\ a_{31} & a_{32} & a_{33} \end{pmatrix} \begin{pmatrix} x_1 \\ x_2 \\ x_3 \end{pmatrix} = \begin{pmatrix} a_{11}x_1 + a_{12}x_2 + a_{13}x_3 \\ a_{21}x_1 + a_{22}x_2 + a_{23}x_3 \\ a_{31}x_1 + a_{32}x_2 + a_{33}x_3 \end{pmatrix} \quad (2.23)$$

Here $a_{ij}$ denotes the element at the intersection of the $i$th row and the $j$th column of $A$, and $x_i$ is the $i$th element of $x$. Note that we can also view $Ax$ as a linear combination of the columns of $A$: compare (2.16) and (2.20).

We've presented quite a few definitions. Now we'll look at a concrete example which should help to illustrate the definitions and indicate why vectors and matrices are so useful.

### Analysing test marks

A group of five students takes five tests, obtaining the marks (out of 20) given in Table 2.2. To analyse the results we first collect the data into a $5 \times 5$ matrix:

$$A = \begin{pmatrix} 15 & 18 & 9 & 12 & 10 \\ 7 & 12 & 6 & 11 & 9 \\ 17 & 16 & 10 & 15 & 16 \\ 10 & 12 & 8 & 10 & 12 \\ 8 & 11 & 5 & 9 & 10 \end{pmatrix} \quad (2.24)$$

We define also the vector of 'ones',

$$e = (1 \quad 1 \quad 1 \quad 1 \quad 1)^{\mathrm{T}} \quad (2.25)$$

*Table 2.2.* Test results for five students.

| Student | Test 1 | Test 2 | Test 3 | Test 4 | Test 5 |
|---------|--------|--------|--------|--------|--------|
| Anderson | 15 | 18 | 9 | 12 | 10 |
| Hughes | 7 | 12 | 6 | 11 | 9 |
| Jones | 17 | 16 | 10 | 15 | 16 |
| Smith | 10 | 12 | 8 | 10 | 12 |
| Wilson | 8 | 11 | 5 | 9 | 10 |

Then we can compute various statistics of interest using matrix–vector multiplication. The vector $Ae$ contains the total marks of each student, since its $i$th component is the inner product of the $i$th row of $A$ with $e$, which is the sum of the elements in the $i$th row of $A$. We multiply by the scalar $\frac{1}{5}$, so that $\frac{1}{5}Ae$ gives the average mark of each student over all the tests.

Similarly, by taking inner products of $e$ with the columns of $A$,

$$e^{\mathrm{T}}A = (1 \quad 1 \quad 1 \quad 1 \quad 1) \begin{pmatrix} 15 & 18 & 9 & 12 & 10 \\ 7 & 12 & 6 & 11 & 9 \\ 17 & 16 & 10 & 15 & 16 \\ 10 & 12 & 8 & 10 & 12 \\ 8 & 11 & 5 & 9 & 10 \end{pmatrix}$$

$$= (57 \quad 69 \quad 38 \quad 57 \quad 57) \tag{2.26}$$

we obtain the total marks for each test.

We can pick out the results for a particular student, or test, using vectors $e_j$, where $e_j$ has '1' in its $j$th position and zeros everywhere else. Thus, for example, Smith's marks are given by

$$e_4^{\mathrm{T}}A = (0 \quad 0 \quad 0 \quad 1 \quad 0)A = (10 \quad 12 \quad 8 \quad 10 \quad 12) \tag{2.27}$$

A glance at Table 2.2, or the matrix $A$, shows that all the students performed relatively poorly on Test 3. We might decide that this test was too hard, and scale the marks for it by some factor $f > 1$, chosen so that (for example) the average mark on Test 3 is equal to the average of the marks for all the other tests. The average marks for the tests are given by $\frac{1}{5}$ times the row vector $e^{\mathrm{T}}A$ in (2.26); thus the average mark for Test 3 is $\frac{1}{5}(e^{\mathrm{T}}A)e_3 = 38/5$. The average of the marks for the other tests is

$$\frac{1}{4 \times 5}(e^{\mathrm{T}}A)\begin{pmatrix} 1 \\ 1 \\ 0 \\ 1 \\ 1 \end{pmatrix} = \frac{1}{4 \times 5}(57 + 69 + 57 + 57)$$

$$= \frac{240}{20} = 12 \tag{2.28}$$

Therefore the scale factor $f$ is given by $(38/5)f = 12$, which gives $f = 60/38 = 30/19$. To carry out the scaling we multiply the third column of $A$ by $30/19$.

Finally, to indicate that scaling exam marks is a tricky business, let's see how we could scale the marks for each test so as to give each student any mark he or she chooses! Let the scale factor for Test $i$ be $x_i$, and let $b_i$ be the total mark desired by the $i$th student. Then $x$ is the solution of the $5 \times 5$ system of equations $Ax = b$, which can be solved by the method described below. (Can you see a flaw in this argument? Consider the likely case $b_1 = b_2 = b_3 = b_4 = b_5 = 100$!)

### Gaussian elimination

Gaussian elimination (GE) is a method for solving a system of $n$ linear equations in $n$ unknowns, where $n$ can be as large as we please. It is precisely the method we used at the start of this chapter to solve equations (2.3)–(2.5). It has a long history – a variant of it for solving systems of three equations in three unknowns appears in the classic Chinese work *Chiu-chang Suan-shu* ('Nine Chapters on the Mathematical Art'), written around 250 BC.

The fundamental idea in GE is a very simple one: we want to reduce a problem that we can't solve immediately to one that we can. Usually, in a general system $Ax = b$ each equation will contain more than one unknown, and so we can't find any of the unknowns directly. We therefore need to look for a special form of $A$ that will allow direct solution. Consider equations (2.3), (2.6) and (2.8), which can be written in matrix notation as

$$\begin{pmatrix} 1 & 1 & 1 \\ 0 & 1 & 3 \\ 0 & 0 & 2 \end{pmatrix} \begin{pmatrix} c \\ b \\ a \end{pmatrix} = \begin{pmatrix} 3 \\ 2 \\ -6 \end{pmatrix} \qquad (2.29)$$

It was these equations that we used to compute $a$, $b$ and $c$. The last, equation (2.8), involves only $a$ and gives us $a = -3$. We substitute this value into the second equation, in which $b$ is then the only unknown, and solve for $b$. Finally, $c$ is obtained from the first equation. The key property of the coefficient matrix in equation (2.29) is that all the elements below the diagonal (top left to bottom right) are zero. Such a matrix is called *upper-triangular*, because the non-zero elements occur in a triangle in the upper right portion of the matrix. It's clear, just from looking at our $3 \times 3$ example, that if $A$ is an $n \times n$ upper-triangular matrix then we can solve $Ax = b$

by computing the unknowns in the order $x_n, x_{n-1}, \ldots, x_1$. For example, if $n = 4$ we can solve

$$
\begin{pmatrix}
t_{11} & t_{12} & t_{13} & t_{14} \\
0 & t_{22} & t_{23} & t_{24} \\
0 & 0 & t_{33} & t_{34} \\
0 & 0 & 0 & t_{44}
\end{pmatrix}
\begin{pmatrix}
x_1 \\
x_2 \\
x_3 \\
x_4
\end{pmatrix}
=
\begin{pmatrix}
b_1 \\
b_2 \\
b_3 \\
b_4
\end{pmatrix}
\tag{2.30}
$$

by computing

$$
\begin{aligned}
x_4 &= b_4/t_{44} \\
x_3 &= (b_3 - t_{34}x_4)/t_{33} \\
x_2 &= (b_2 - t_{23}x_3 - t_{24}x_4)/t_{22} \\
x_1 &= (b_1 - t_{12}x_2 - t_{13}x_3 - t_{14}x_4)/t_{11}
\end{aligned}
\tag{2.31}
$$

This process is called *back-substitution*: once an unknown is found it is substituted back into the previous equation.

The method of Gaussian elimination reduces the equations $Ax = b$ to a system with the same solution and with an upper-triangular coefficient matrix. The triangular system is then solved by back-substitution. The reduction is done by repeatedly using one equation to eliminate variables from the remaining equations. We'll illustrate the process for the $4 \times 4$ system

$$
Ax =
\begin{pmatrix}
2 & 1 & 0 & 4 \\
-6 & -3 & 1 & -2 \\
2 & 5 & 4 & 5 \\
4 & 1 & -2 & 10
\end{pmatrix}
\begin{pmatrix}
x_1 \\
x_2 \\
x_3 \\
x_4
\end{pmatrix}
=
\begin{pmatrix}
4 \\
1 \\
25 \\
2
\end{pmatrix}
\tag{2.32}
$$

There are three main stages to the elimination. First, we eliminate $x_1$ from the second, third and fourth equations by adding to them multiples of 3, $-1$ and $-2$ of the first equation, obtaining

$$
\begin{pmatrix}
2 & 1 & 0 & 4 \\
0 & 0 & 1 & 10 \\
0 & 4 & 4 & 1 \\
0 & -1 & -2 & 2
\end{pmatrix}
\begin{pmatrix}
x_1 \\
x_2 \\
x_3 \\
x_4
\end{pmatrix}
=
\begin{pmatrix}
4 \\
13 \\
21 \\
-6
\end{pmatrix}
\tag{2.33}
$$

The element '2' in the $(1, 1)$ position in $A$ is called the *pivot element* for this first elimination step. The numbers 3, $-1$ and $-2$ that gave the required multiples of the first equation are called *multipliers*.

The first column of the matrix now has the form required of an upper-triangular matrix, and we can ignore the first row and column and concentrate on reducing the remaining $3 \times 3$ subsystem to triangular form. However, immediately we hit a problem – our pivot for the next stage, the $(2, 2)$ element, is zero. The second equation doesn't contain $x_2$ and so can't be used to eliminate this unknown from the remaining equations. A simple way round this difficulty is to swap the second and third equations, bringing the element '4', which is currently in the $(3, 2)$ position, to the pivot position. This gives

$$\begin{pmatrix} 2 & 1 & 0 & 4 \\ 0 & 4 & 4 & 1 \\ 0 & 0 & 1 & 10 \\ 0 & -1 & -2 & 2 \end{pmatrix} \begin{pmatrix} x_1 \\ x_2 \\ x_3 \\ x_4 \end{pmatrix} = \begin{pmatrix} 4 \\ 21 \\ 13 \\ -6 \end{pmatrix} \tag{2.34}$$

To complete the second stage of the elimination we just add $\frac{1}{4}$ times the second equation to the fourth, getting

$$\begin{pmatrix} 2 & 1 & 0 & 4 \\ 0 & 4 & 4 & 1 \\ 0 & 0 & 1 & 10 \\ 0 & 0 & -1 & \frac{9}{4} \end{pmatrix} \begin{pmatrix} x_1 \\ x_2 \\ x_3 \\ x_4 \end{pmatrix} = \begin{pmatrix} 4 \\ 21 \\ 13 \\ -\frac{3}{4} \end{pmatrix} \tag{2.35}$$

(Notice that the zero pivot actually reduced our work because it meant that after the interchange, the matrix already had a zero in the $(3, 2)$ position.) We now add the third equation to the fourth, to obtain the upper-triangular system

$$\begin{pmatrix} 2 & 1 & 0 & 4 \\ 0 & 4 & 4 & 1 \\ 0 & 0 & 1 & 10 \\ 0 & 0 & 0 & \frac{49}{4} \end{pmatrix} \begin{pmatrix} x_1 \\ x_2 \\ x_3 \\ x_4 \end{pmatrix} = \begin{pmatrix} 4 \\ 21 \\ 13 \\ \frac{49}{4} \end{pmatrix} \tag{2.36}$$

Using back-substitution (equations (2.31)), we find that

$$x^T = (-1 \quad 2 \quad 3 \quad 1) \tag{2.37}$$

which, it is easy to confirm, is the solution to the system (2.32).

In practice it is quite common to have to solve several linear systems with the *same* matrix $A$, but *different* right-hand sides. This might arise, for example, if the experiment that yielded the data given in Table 2.1 were repeated with the same time intervals but with different chemicals. An important feature of GE is that after one system $Ax = b$ has been solved, a further system with a different right-hand side can be solved with relatively little extra effort. The reason is that the elimination operations, and hence the final upper-triangular matrix, depend only on $A$. To process a new right-hand side, $c$, we simply carry out the same elimination operations on $c$ that were applied to $b$.

For example, to solve equations (2.3)–(2.5) with the right-hand side replaced by $(3\ 7\ 13)^T$, we mimic the operations used to obtain (2.6)–(2.8). Thus

$$
\begin{pmatrix} 3 \\ 7 \\ 13 \end{pmatrix} \quad \begin{matrix} \text{subtract} \\ \text{row 1 from} \\ \text{rows 2 and 3} \end{matrix} \to \begin{pmatrix} 3 \\ 4 \\ 10 \end{pmatrix} \quad \begin{matrix} \text{subtract} \\ \text{twice row 2} \\ \text{from row 3} \end{matrix} \to \begin{pmatrix} 3 \\ 4 \\ 2 \end{pmatrix} \quad (2.38)
$$

We can then solve the triangular system (compare equation (2.29))

$$
\begin{pmatrix} 1 & 1 & 1 \\ 0 & 1 & 3 \\ 0 & 0 & 2 \end{pmatrix} \begin{pmatrix} c \\ b \\ a \end{pmatrix} = \begin{pmatrix} 3 \\ 4 \\ 2 \end{pmatrix} \quad (2.39)
$$

This gives $a = b = c = 1$.

### Computational cost

Our examples of GE have so far been for problems with $n = 3$ or 4, where the computational effort is small. In practical problems $n$ may be very large, and it is important to have a means for assessing the computational cost of GE. For example, we may want to estimate the time required to solve a $100 \times 100$ system of equations on a particular computer. A useful way to quantify the amount of work required by GE is to count the number of arithmetic operations. In doing this, for simplicity, we shall regard a division as being equivalent to a multiplication, and we shall ignore subtractions and additions.

In the first stage of GE on an $n \times n$ matrix we need $n - 1$ divisions to compute the multipliers, and then $(n - 1) \times n$ multiplications to carry out the operations of adding the first equation to each of the $n - 1$ remaining equations. (Note that we don't need to compute the elements that we know are going to become zero!) This gives a total of $n - 1 + n^2 - n = n^2 - 1$ multiplications for the first stage. The second stage is like the first, but this time the matrix has dimension $n - 1$, so its cost is $(n - 1)^2 - 1$ multiplications, and likewise for the remaining elimination stages. The total cost of reducing the system to upper-triangular form is therefore

$$(n^2 - 1) + [(n - 1)^2 - 1] + \cdots + (2^2 - 1) \qquad (2.40)$$

multiplications, which we can write using the standard formula as*

$$\sum_{k=1}^{n} (k^2 - 1) = \frac{1}{6} n(2n + 1)(n + 1) - n$$

$$= \frac{2n^3 + 3n^2 - 5n}{6} \approx \frac{n^3}{3} \qquad (2.41)$$

Since we are interested in the operation count mainly for large values of $n$, we have retained only the fastest-growing term in the last line, reasoning that $n$ and $n^2$ will be small compared with $n^3$ when $n$ is large. (Incidentally, note that

$$\int_0^n x^2 \, dx = \frac{n^3}{3} \approx \sum_{k=1}^{n} k^2 \qquad (2.42)$$

In general, if $n$ is large, for powers other than 2 it is valid to use the corresponding approximation of a sum by an integral.)

The back-substitution stage requires a total of

$$1 + 2 + \cdots + n = \sum_{k=1}^{n} k = \frac{n(n + 1)}{2} \approx \frac{n^2}{2} \qquad (2.43)$$

multiplications (the pattern can be seen in equations (2.31)). Overall, then, GE solves an $n \times n$ system of equations in approximately $n^3/3$ multiplications.

To solve a system with the same matrix and a different right-hand side requires an additional $n^2/2$ multiplications for the elimination stage plus $n^2/2$ for the back-substitution, making a total of $n^2$ extra multiplications. If $n$ is not particularly small (bigger than 10, say) then $n^2$ is much less than $n^3/3$ and we see that it is much more efficient to re-use the 'old' multipliers and the triangular matrix than to start the elimination again from the beginning.

---

* The summation sign $\Sigma$ indicates that a number of terms are to be added together. For example, $\Sigma_{k=1}^{3} k$ gives $1 + 2 + 3 = 6$, while $\Sigma_{k=3}^{6} k^2$ gives $3^2 + 4^2 + 5^2 + 6^2 = 86$.

But is GE the *most* efficient way to solve general linear systems? The theoretical answer is that it is not. Mathematicians have constructed algorithms that can solve an $n \times n$ system of equations using a number of multiplications that is proportional to $n^\alpha$, where $\alpha$ is a constant lying between 2 and 3. Thus for large enough $n$ these algorithms are more efficient than GE. Over the last couple of decades the value of $\alpha$ has been reduced many times, as sharper techniques for the construction have been found. Impressive as they may be from the theoretical point of view, these results have had little impact on practical computation. The main reason is that for most of the 'fast' methods the constants of proportionality are such that $n$ has to be extremely large before any significant computational advantage is obtained over GE – and such large systems are invariably of a special form (see p. 54) for which these fast algorithms are no longer the most efficient.

## Computer solution of linear equations

Solving systems of linear equations is the basic form of calculation in scientific computation. For although mathematical models are usually non-linear, most methods for solving non-linear problems involve some form of linearization, which leads to systems of linear equations. Computer programs implementing GE are now available in most scientific program libraries, and are usually among the most heavily used routines. While the amateur cannot hope to produce programs of a similar quality, it is nevertheless easy to write a straightforward implementation of GE.

Here is an example of such a program, written in BASIC, an easy-to-learn programming language taught in many schools and available on most microcomputers. The program has purposely been kept as simple as possible. Since it doesn't incorporate row interchanges, it can fail with division by zero at line 100, and it may return computed answers that are less accurate than they might otherwise have been (see p. 52). Nevertheless, it indicates the essential simplicity and elegance of GE. In the program, N is the number of equations, and the arrays $A(I,J)$, $X(I)$ and $B(I)$ hold the elements of the matrix $A$ and the vectors $x$ and $b$ respectively.

```
10 REM Program to solve Ax=b by Gaussian Elimination
20 READ N
30 DIM A(N,N), X(N), B(N)
40 FOR J = 1 TO N: FOR I = 1 TO N: READ A(I,J):
   NEXT I: NEXT J
```

```
 50 FOR I = 1 TO N: READ B(I): NEXT I
 60 REM The following data are for equations (1),
       (2),(3), and (15)
 70 DATA 3, 1,1,1, 1,2,3, 1,4,9, 3,5,1
 80 REM Reduction to upper-triangular form
 90 FOR K = 1 TO N
100   FOR I = K + 1 TO N: M = A(I,K)/A(K,K)
110     FOR J = K + 1 TO N: A(I,J) = A(I,J) - M *
        A(K,J): NEXT J
120     B(I) = B(I) - M * B(K)
130   NEXT I
140 NEXT K
150 REM Back-substitution
160 X(N) = B(N)/A(N,N)
170 FOR I = N - 1 TO 1 STEP -1
180   S = 0: FOR J = I + 1 TO N: S = S + A(I,J) * X(J):
      NEXT J
190   X(I) = (B(I) - S)/A(I,I)
200 NEXT I
210 REM Print answer
220 FOR I = 1 TO N: PRINT X(I): NEXT I
```

If you have access to a microcomputer, you could use this program as the basis for a variety of experiments. Some ideas are as follows.

(1) Run the program for various values of $n$ (e.g. $n = 5, 10, 15, 20, 25$) and see if the execution time $t_n$ is proportional to $n^3$, as would be predicted from the operation count. (It doesn't really matter how you choose the arrays A and B; a convenient way is to assign to their elements random numbers from BASIC's random number generator.) Plot a graph of $t_n$ against $n^3$ (using your microcomputer's graphics, if available) and look for straight-line behaviour.

(2) Modify the program to incorporate partial pivoting (as described on p. 51).

(3) Extend the program so that it tests how well the computed solution $\hat{x}$ satisfies the equations, by calculating the residual vector $r = b - A\hat{x}$. Since the arrays A and B are changed by the elimination routine, you will have to make a copy of them at the start of the routine. In particular, you might like to investigate the size of the performance measure $\rho = |r|/(u|A|\,|\hat{x}|)$, where $|r|$ and $|A|$ denote the largest elements in absolute value of the vector $r$ and the matrix $A$, and $u$ is the rounding unit (see p. 49). Ideally, $\rho$ should be of the order of 1, and it almost always is if partial pivoting is used. For $n = 4$, can you find a matrix $A$ and

a vector $b$ such that the GE program (without pivoting) gives a value $\rho > 100$?

(4) For large $n$, the program spends most of its time executing the statement inside the loop in line 110, since this is the innermost of the three nested loops. This statement involves a multiplication, an addition, and some subscripting of two-dimensional arrays. (Subscripting is the operation in which the computer works out where an array element is stored in computer memory; for example A(N,N) will probably be stored $N^2$ 'locations' along from A(1,1).) Investigate the relative costs of these operations on your computer, by timing the execution of statements like T = A(I,J), T = R * S. (In interpreted BASIC, subscripting is usually about as expensive as multiplication!)

It is interesting to compare the performance of a microcomputer with that of a mainframe machine for the speed of solving linear equations. Some timings for 100 equations in 100 unknowns are given in Table 2.3; these were obtained using a GE routine from the LINPACK library of FORTRAN programs for solving linear equations. (For the BBC machine a BASIC translation of the FORTRAN GE routine was used, and the timings were extrapolated from actual timings for $n = 60$, the largest value of $n$ possible with the BBC's limited memory). As can be seen, in this application the BBC and IBM microcomputers are orders of magnitude slower than the mainframes, but the difference is perhaps not as great as might be expected.

Surprisingly, on some machines (particularly the supercomputers) minor modifications to the way GE is programmed can have a dramatic effect on the execution time. An active topic of research in recent years has been how to implement GE so that it performs as efficiently as possible on particular classes of computer.

*Table 2.3.* Time taken to solve $Ax = b$ for $n = 100$ on different computers.

| Computer (Language) | Time | |
|---|---|---|
| BBC microcomputer (BASIC) | 36 | minutes |
| BBC microcomputer (BASIC + machine code) | 13 | minutes |
| IBM PC with 8087 co-processor (FORTRAN) | 59 | seconds |
| VAX 11/780 (FORTRAN) | 4.96 | seconds |
| CDC Cyber 205 (FORTRAN) | 0.082 | seconds |
| Cray X-MP-1 (FORTRAN) | 0.032 | seconds |

### Rounding errors

An intrinsic limitation of computer arithmetic is that numbers are stored only to a finite number of digits, $t$ say. Consider a computer which works in base 10 (in practice, base 2 is the most common). A number that has an infinite decimal expansion, such as $x = \frac{1}{3} = 0.333\ldots$, cannot be represented exactly in the computer; the best the computer can do is to approximate $x$ by $\hat{x} = 0.33\ldots3$ ($t$ digits). The difference $x - \hat{x}$ is called a *rounding error*. Rounding errors also occur when the computer attempts to perform arithmetic operations. For example, if two $t$-digit machine numbers $y$ and $z$ are multiplied together, then the exact product will have $2t$ or $2t - 1$ digits; since the machine can store only $t$ digits, those beyond the $t$th place in the product $yz$ will be 'lost', and the computed product will contain a rounding error. In general, each fundamental arithmetic operation $(\times, \div, +, -)$ is subject to a rounding error when carried out in the computer's finite-precision ($t$-digit) arithmetic.

The size of a rounding error depends on the size of the number with which it is associated. The maximum possible rounding error in numbers of order 1, called the *rounding unit*, varies from computer to computer, and is usually between $10^{-6}$ and $10^{-15}$. For our hypothetical base 10, $t$-digit machine, the rounding unit would be $\frac{1}{2}10^{1-t}$. The following BASIC program estimates the rounding unit. It finds the largest negative power of 2 which, when added to 1 in the computer arithmetic, gives back the value 1.

```
10 X = 1
20 X = X/2
30 Y = X + 1
40 IF Y > 1 THEN 20
50 PRINT "Rounding unit is approximately"; X
```

Although an individual rounding error is tiny, the sum of many of them can be large if there is little cancellation in the sum. For example, there are tens of thousands of operations in solving a $50 \times 50$ system of equations by GE, and if the rounding errors in these operations were to accumulate, the accuracy of the computed solution could be severely affected. Indeed, in the 1940s, when computer solution of linear equations was first contemplated, there was widespread feeling that, if $n$ was at all large, rounding errors would swamp the computed solution, rendering it totally inaccurate. We now know that as long as certain precautions are taken, the rounding errors are relatively harmless.

To see the possible ill-effects of rounding errors, consider the system of equations

$$\begin{pmatrix} 0.0001 & 1 \\ 1 & 1 \end{pmatrix} \begin{pmatrix} x_1 \\ x_2 \end{pmatrix} = \begin{pmatrix} 1 \\ 0 \end{pmatrix} \tag{2.44}$$

Performing GE exactly, we subtract 10 000 times the first equation from the second, getting

$$\begin{pmatrix} 0.0001 & 1 \\ 0 & -9999 \end{pmatrix} \begin{pmatrix} x_1 \\ x_2 \end{pmatrix} = \begin{pmatrix} 1 \\ -10\,000 \end{pmatrix} \tag{2.45}$$

so that

$$x_2 = \frac{10\,000}{9999}$$

$$x_1 = 10\,000 \left( 1 - \frac{10\,000}{9999} \right) = -\frac{10\,000}{9999} \tag{2.46}$$

If, however, we perform the calculation in three-digit arithmetic, the triangular system is

$$\begin{pmatrix} 0.0001 & 1 \\ 0 & -10\,000 \end{pmatrix} \begin{pmatrix} x_1 \\ x_2 \end{pmatrix} = \begin{pmatrix} 1 \\ -10\,000 \end{pmatrix} \tag{2.47}$$

($-9999$ has four digits and has to be rounded down to $-10\,000$, which has just one significant digit and so is indeed a 'three-digit number'). The computed solution is, therefore,

$$\hat{x}_2 = 1, \qquad \hat{x}_1 = 0 \tag{2.48}$$

While $\hat{x}_2$ differs only slightly from $x_2$, it is clear that $\hat{x}_1$ is completely the wrong answer!

The reason for this poor performance is that the small pivot, 0.0001, caused us to subtract a relatively large multiple of one equation from the other, and this swamped the $(2, 2)$ element of $A$. In fact, if we'd started instead with the system

$$\begin{pmatrix} 0.0001 & 1 \\ 1 & 0 \end{pmatrix} \begin{pmatrix} x_1 \\ x_2 \end{pmatrix} = \begin{pmatrix} 1 \\ 0 \end{pmatrix} \tag{2.49}$$

where the $(2, 2)$ element '1' of $A$ has been replaced by '0', then we would have obtained the *same* final upper-triangular system! Clearly, because of the small pivot, the numerical procedure is relatively indifferent to the value of the $(2, 2)$ element of $A$, whereas the mathematical problem is certainly not.

The phenomenon displayed in this $2 \times 2$ example is called *instability*. Thus, GE in its basic form is unstable: in finite-precision arithmetic it can produce a computed solution that is much less accurate than we would reasonably expect.

Fortunately, there is a simple way to 'stabilize' the method. In our $2 \times 2$ example the root of the difficulty was the small pivot $a_{11} = 0.0001$, which produced a large multiplier $a_{21}/a_{11} = 1/0.0001 = 10\,000$. Suppose we interchange the two equations:

$$\begin{pmatrix} 1 & 1 \\ 0.0001 & 1 \end{pmatrix} \begin{pmatrix} x_1 \\ x_2 \end{pmatrix} = \begin{pmatrix} 0 \\ 1 \end{pmatrix} \qquad (2.50)$$

The pivot is now 1 and the multiplier 0.0001, and in the same three-digit arithmetic GE produces the triangular system

$$\begin{pmatrix} 1 & 1 \\ 0 & 1 \end{pmatrix} \begin{pmatrix} x_1 \\ x_2 \end{pmatrix} = \begin{pmatrix} 0 \\ 1 \end{pmatrix} \qquad (2.51)$$

since 0.9999 rounds to 1. The computed solution is therefore

$$\hat{x}_2 = 1, \qquad \hat{x}_1 = -1 \qquad (2.52)$$

which is as accurate as we could expect from a three-digit computation.

The computation just described is an example of *GE with partial pivoting*. In this version of GE we examine, at each stage $k$, all the elements below the diagonal in the pivot column (the $k$th column) to find the one that is largest in magnitude. If this largest element is in the $r$th row, then we interchange equations $k$ and $r$, thus bringing the largest of the potential pivots into the pivot position. For example, for the system (2.32) (p. 42) we would interchange the first two equations at the first stage, to bring the largest element in the first column, '$-6$', into the pivot position. Partial pivoting ensures that every multiplier is bounded in magnitude by 1, so that we shall never find ourselves adding a large multiple of the pivot row to a later row.

As we have mentioned, in the early years of computing the effect of rounding errors on GE was not well-understood. But pioneering research by J. H. Wilkinson in the 1950s and early 1960s showed that any doubts about the numerical stability of GE with partial pivoting were unfounded. Wilkinson developed a rounding error analysis which showed the following result:

> When a system of linear equations is solved in finite-precision arithmetic by GE with partial pivoting, the computed solution $\hat{x}$ is the exact solution of a

perturbed system of equations $(A + E)\hat{x} = b$, where
the elements of the matrix $E$ are usually about the size
of rounding errors in the elements of $A$.

(As an example, $\hat{x}$ in equations (2.52) is the exact solution of
equations (2.51), which are precisely equations (2.50) with $-0.0001$
added to the $(2, 1)$ element.) This is a very satisfactory result, because
it says that the effect of the rounding errors on GE is no worse than
the effect of simply storing $A$ on the computer (that is, converting the
elements of $A$ to $t$-digit machine numbers). The qualifier 'usually' in
Wilkinson's statement of his result relates to certain cases where the
elements of $E$ can be rather large if $n$ is large, but these cases are
extremely rare and can be ignored for practical purposes.

In summary, GE with partial pivoting is a very stable algorithm,
and it is the method of choice for solving general systems of linear
equations on a computer.

### Ill-conditioning

Unfortunately, Wilkinson's result does not necessarily imply that $\hat{x}$
is a good approximation to $x$. To see this, look at the equations

$$\begin{pmatrix} 1 & 1 \\ 1 & 1 + \varepsilon \end{pmatrix} \begin{pmatrix} x_1 \\ x_2 \end{pmatrix} = \begin{pmatrix} 1 \\ 2 \end{pmatrix} \tag{2.53}$$

whose solution is

$$x_1 = 1 - \frac{1}{\varepsilon}, \qquad x_2 = \frac{1}{\varepsilon} \tag{2.54}$$

If $\varepsilon$ changes from 1 to 0.01, which causes a very small change in the
matrix, then $x_1$ and $x_2$ change by about 100! Geometrically, we are
trying to find the intersection of two almost parallel lines; a small
movement of one line can move the point of intersection a long way.
This phenomenon, whereby a small change in $A$ or $b$ can produce a
much larger change in $x$, is called *ill-conditioning*. In practice, if a
system of equations is ill-conditioned then usually we can't find $x$ to
very high accuracy. Of course, there are various degrees of ill-
conditioning. One extreme is when the equations have no solution!
If we set $\varepsilon = 0$ then this is the case for equations (2.53), because
$x_1 + x_2$ can't simultaneously equal both 1 and 2; this is indicated by
the solution (2.54) 'blowing up' as $\varepsilon$ tends to 0, and by the matrix in
equations (2.53) having equal columns for $\varepsilon = 0$.

### Use and abuse of the inverse matrix

Consider the trivial linear equation $ax = b$, where $a \neq 0$ and $b$ are scalars. The natural way to solve it is to write $x = b/a$. A more formal statement of this approach is that both sides of the equation are multiplied by $a^{-1}$ to get $x = a^{-1}b$. Here $a^{-1}$ denotes the number such that $a^{-1}a = 1$, which is called the *inverse* of $a$. Clearly, $a^{-1} = 1/a$. The notion of 'inverse' can be generalized to an $n \times n$ matrix $A$. By analogy with the scalar case, we would like to solve the system of equations $Ax = b$ by pre-multiplying by an $n \times n$ matrix $A^{-1}$ to obtain

$$x = A^{-1}Ax = A^{-1}b \qquad (2.55)$$

The key property required of $A^{-1}$ is that $x = A^{-1}Ax$, which will be true if $A^{-1}A$ is the matrix analogue of 1 for scalars, namely the identity matrix $I$. The identity matrix is the matrix with ones on the diagonal and zeros everywhere else. For $n = 3$ it is

$$I = \begin{pmatrix} 1 & 0 & 0 \\ 0 & 1 & 0 \\ 0 & 0 & 1 \end{pmatrix} \qquad (2.56)$$

It is easy to see that for any vector $x$, $Ix = x$. Thus we can *define* $A^{-1}$ as that matrix for which $A^{-1}A = I$ (provided such a matrix exists). You may be familiar with the following useful formula which gives the inverse of a $2 \times 2$ matrix:

$$\begin{pmatrix} a & b \\ c & d \end{pmatrix} = \frac{1}{ad - bc} \begin{pmatrix} d & -b \\ -c & a \end{pmatrix} \qquad (2.57)$$

Because of the simplicity and elegance of the formula $x = A^{-1}b$, it is tempting to try to use it to solve systems of equations as follows: compute $A^{-1}$, and then form its product with $b$. While this approach is convenient for $n = 2$, when we are probably working in exact arithmetic, it is not an expedient computational approach in general. For example, to solve $3x = 24$, in four-digit arithmetic say, the natural approach is to carry out the trivial back-substitution $x = 24/3 = 8$, obtaining the exact answer with one division. The $A^{-1}b$ approach would evaluate $3^{-1} = 0.3333$ (to four digits) and then form $x = 0.3333 \times 24 = 7.999$. At the cost of an extra multiplication we've obtained a less accurate answer!

In the $n \times n$ case the assessment goes as follows. For general $A$, the most efficient way to compute $A^{-1}$ is to use GE to solve the linear systems $Ax_i = e_i$, $i = 1, \ldots, n$, where $x_i$ and $e_i$ are the $i$th columns of $A$ and $I$ respectively (these equations come from equating columns in the matrix equation $AA^{-1} = I$). The cost of solving these systems turns out to be $n^3$ multiplications, if we take advantage of two convenient properties: each system has the same coefficient matrix (see p. 33), and the right-hand sides have a regular pattern of zero elements.

Recall, however, that GE solves $Ax = b$ in $n^3/3$ multiplications. The 'inverse' method therefore triples the computational work; also it can be shown to give no greater accuracy than GE. The $A^{-1}b$ approach is clearly not to be recommended.

Despite the great utility of the matrix inverse in the *analysis* of linear algebra problems, it is a maxim of matrix computations that in practice one nearly always can – and wherever possible *should* – avoid computing matrix inverses. It is particularly important to heed this advice when $n$ is large, as we shall now explain.

In practice, large matrices ($n \geqslant 100$, say) are usually *sparse* – that is, a large proportion of their elements are zero. Sparsity is intrinsic to many mathematical models. For example, in modelling the forces on a bridge, equations will be developed that express the balance of forces along the supports at each 'node'. In a long bridge only a few of the many supports will meet at any particular node, and thus each equation will involve only a few unknowns. Therefore in the associated linear system $Ax = b$, each row of $A$ will have only a few non-zero elements. Sparsity can be expected to be present in a model of any physical problem that has such a 'local influence' principle.

Special cases of sparse matrices are *banded matrices*, in which the non-zero elements occur in a band straddling the main diagonal. The most frequently occurring type of banded matrix is the *tridiagonal matrix* (three diagonals). A well-known tridiagonal matrix is the *second-difference* matrix, illustrated for $n = 5$ by

$$A = \begin{pmatrix} -2 & 1 & 0 & 0 & 0 \\ 1 & -2 & 1 & 0 & 0 \\ 0 & 1 & -2 & 1 & 0 \\ 0 & 0 & 1 & -2 & 1 \\ 0 & 0 & 0 & 1 & -2 \end{pmatrix} \qquad (2.58)$$

There are two features of sparse matrices that are particularly relevant to our discussion. The first is that, if $A$ is a large, sparse

$n \times n$ matrix, then GE for solving $Ax = b$ can usually be arranged so that it requires appreciably less computational resources than the $n^3/3$ multiplications and $n^2$ storage locations required when $A$ is full. The operation count, and the computer storage required, are typically proportional to the number of non-zero elements, which is often a small fraction of $n$. This is easily seen for a tridiagonal matrix: most of the elements in the lower triangle are already zero, and the elimination operations do not disturb these zeros, so GE has relatively little work to do. In fact, GE requires just $5n$ multiplications to solve an $n \times n$ tridiagonal system of equations.

Of course, the non-zero elements in a sparse matrix need not occur in a regimented pattern; they may be scattered about the matrix in an apparently random fashion. But even in these cases GE can be adapted to take advantage of the sparsity. The essential idea is to interchange rows and columns at each stage of the elimination, not only to achieve stability, which is the aim of partial pivoting, but also to limit the amount of *fill-in*, which happens when an element which is zero at one stage becomes non-zero at the next. How to organize GE for sparse matrices is a topic of much current research.

The second important feature of sparse matrices is that their inverses are usually *not* sparse. For example, the inverse of the second-difference matrix (2.58) is the full matrix

$$\begin{pmatrix} \frac{5}{6} & \frac{2}{3} & \frac{1}{2} & \frac{1}{3} & \frac{1}{6} \\ \frac{2}{3} & \frac{4}{3} & 1 & \frac{2}{3} & \frac{1}{3} \\ \frac{1}{2} & 1 & \frac{3}{2} & 1 & \frac{1}{2} \\ \frac{1}{3} & \frac{2}{3} & 1 & \frac{4}{3} & \frac{2}{3} \\ \frac{1}{6} & \frac{1}{3} & \frac{1}{2} & \frac{2}{3} & \frac{5}{6} \end{pmatrix} \qquad (2.59)$$

(Notice the beautiful symmetries in the elements; this is typical of the inverses of tridiagonal matrices.) Thus, to solve $Ax = b$ by computing $A^{-1}$ and then $x = A^{-1}b$ requires of the order of $n^2$ multiplications. Moreover, if $n$ is large, then it might not even be possible to store the $n^2$ elements of $A^{-1}$ in the computer!

## The future

One of the reasons for the importance of GE in modern matrix computations is its versatility. It is the method of choice for solving linear equations, whether small and dense, or large and sparse – with possibly millions of equations (such large problems have been solved). And GE is able to take advantage of various special properties of matrices, such as symmetry (when $a_{ij} = a_{ji}$, as in the second-difference matrix (2.58)). Furthermore, it has been found that suitable implementations of GE perform extremely well on many of today's supercomputers. Two decades ago few people would have predicted that GE would be so successful over such a wide range of matrices and computers; it will be interesting to see whether the method retains its eminence in future years, as newer computer architectures are introduced, and new algorithms are developed.

## Further reading

I. Bradley and R. L. Meek, *Matrices and Society* (Penguin, 1986).

G. E. Forsythe, M. A. Malcolm and C. B. Moler, *Computer Methods for Mathematical Computations* (Prentice-Hall, 1977), Chapter 3: 'Linear systems of equations'.

W. W. Sawyer, *Prelude to Mathematics* (Penguin, 1955), Chapter 8: 'Matrix algebra'.

G. Strang, *Linear Algebra and Its Applications*, second edition (Academic Press, 1980).

# 3

## MAKING GUESSES AND
## TAKING RISKS

### ANDREW NOBLE

#### Roll the dice – and win a car!

Wimbledon Village held a fair in 1986 for the first time in many years. All the local shopkeepers, businesses, clubs and societies had displays, entertainments and competitions. One of the most exciting of these competitions was held by a car dealer. For £5 a throw a competitor could roll simultaneously six dice; six 6's would win a car.

As I walked up Wimbledon Hill I asked myself what was my chance of winning with one attempt. Now, for any unbiased regular single die the chance of throwing a 6 is one in six, or approximately 0.167, 16.7%. If I throw that die 10 000 times in ten sets of 1000, *on average* it will land with 6 face up between 160 and 170 times in each 1000 throws. If I now throw two dice together, there are 36 possible results (list them on a piece of paper if you like), but a pair of 6's appears only once. So the chance of throwing a pair of 6's in a single throw of two dice is 1 in $6^2$, or 1 in 36. By continuing this argument it became clear to me that the chance of throwing six 6's in a single throw of six dice simultaneously is 1 in $6^6$, which is 1 in 46 656.

I thought they were rather small odds. After all, millions of people try the football pools each week against very much greater odds than that. This dice throwing was well worth a £5 flutter. Sadly, however, as I walked in to the showroom I was told that the car has just been won. I saved my £5, but I could not win the car either.

How did the car dealer feel, I wonder? Did he make a surplus to give to charity? Had he insured himself against loss? If he had insured himself, what premium had he paid? How did the insurer decide how much the car dealer ought to pay? You might like to think about these questions. Incidentally, at a recent golf

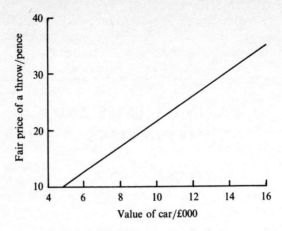

*Figure 3.1*. The 'fair price' of a throw in the win-a-car game
for different car values.

competition in which I played there was a prize of a car for getting
a hole in one at a particular hole. (It is fairly obvious that I have been
obsessed with winning a car.) The insurance premium paid by the
donor of the prize was £50!

If the car was worth £5000, then it seems that the insurer thought
that the chance of a hole in one was rather less than one in a
hundred. (The insurer has to make a profit over all risks, and the
insurance agent takes a commission!) From sad experience, however,
I know that the chance is far smaller than that. The insurer and the
agent are making a very comfortable living on this business.

Even today, £5 is not an insignificant amount of money. The dice
rolling game was for charity, so the aim was to extract as much as
possible from each gambler. Also, the car dealer had almost certainly
insured against a win and so stood to lose only the insurance
premium. But if this had been a straight gamble, what would have
been a 'fair price' for a single throw?

As we have seen, the chance of a 'win' with a single throw is 1 in
46 656. So it would be reasonable to say that a fair price for a throw
is $V/46 656$, where $V$ is the value of the car. It would be reasonable
too to assume that this is the 'on the road' cost to a normal purchaser
although, of course, the cost to the dealer is rather less. (Incidentally,
it might be this latter cost that the dealer quotes to the insurance
company, in order to keep the premium down.)

Now, for cars of various values it is possible to plot the fair price
of a throw against car values, as shown in Fig. 3.1. The car would
have to be worth £233 280 for £5 to be a fair price!

However, most people at Wimbledon Fair thought that £5 was not too unreasonable a flutter for a Metro. This is an example of something frequently observed: most people's sense of fairness or risk is wildly different from what a logical assessment would indicate. This is especially true of unfamiliar circumstances or dangers which are perceived to be catastrophic.

So, as I resigned myself to life without the prize car, another thought came to me about a characteristic of guesses and gambling which people find easy to accept in certain situations but almost impossible to accept in others. The *first* entrant to that competition could have won the prize, however long the odds. Improbable things do happen. Yes, of course they do. Ah, but what about Flixborough, Chernobyl, the *Challenger*, King's Cross, Clapham Junction? These catastrophes were supposed to be impossible, weren't they? No, only highly improbable.

When a complex machine depends for its safety on hundreds or thousands of individual components, even with the cleverest of interlocking safety arrangements there is almost certainly one path to disaster, however unlikely that path may be. The machine could fail within the next minute; on the other hand it might not fail within the next billion minutes (about 2000 years). But what is meant by that? Man-made materials are unlikely to last that long; so let us say we expect the machine to operate by itself, without inspection, for perhaps a week – failure within that time is virtually impossible, isn't it? Then – bang! The broken turbine blade has ruptured the fuel tank; the control room lights wink reassuringly green while the atomic pile's cooling system boils dry; the rubber O-ring fails; the tiny fire leads to a 'flashover'; the ends of two loose wires rub together. And ever present is human error.

All these are examples of the will-o'-the-wisp science of probability – so vital in so many applications to real life, so difficult to comprehend and accept. So much of mankind really only likes and admires a winner. How inconvenient that in so many things we do, even if we do them absolutely correctly, there lurks the chance of failure. 'Hard luck' takes on a deeper meaning. That is why Napoleon required his marshals to be lucky above all else.

## Coins, balls and birthdays

Yet the branch of mathematics known as *probability theory* has come a long way since it struggled against prejudice and narrowness of thought when it was first introduced. Logic, intuition and experience of the physical world are interdependent, sometimes one taking

*Table 3.1.* The possible outcomes of
placing two balls, *a* and *b*, in two
containers, I and II.

| Event | Container | |
|---|---|---|
| | I | II |
| 1 | a b | |
| 2 | | a b |
| 3 | a | b |
| 4 | b | a |

the lead in scientific advance, sometimes another. Probability theory began as this chapter began – with the study of chance. Nowadays newspapers are forever reporting the results of opinion polls, market surveys and voting intentions; statistics of one kind or another seem to dominate our lives. Even if they still smile at the aphorism about 'lies, damned lies and statistics', most people have acquired a feeling for what 'one in six' means, and this is what is called *physical* or *statistical probability*, the subject of this chapter.

It is necessary, too, to realize that probabilistic models of systems are concerned as often with things which cannot be observed as with things which we can see or physically count. The mathematician not only builds conceptual models, but also conducts conceptual experiments. Many of these may be incapable of actual realization, but they still yield important results for real applications. Striking examples are to be found in atomic physics.

The result of an experiment is called an *event*, and the experiment is 'idealized' in the sense that certain outcomes are excluded. For example, in coin-tossing experiments the possibility of the coin standing on edge is not admitted, although in the physical world that could happen. The aggregate of all possible outcomes is called a *sample space*, and each event is a *sample point* or collection of sample points in that sample space. Thus the experiment of tossing a coin once is defined by a sample space of two points, one of which represents the event 'head' and the other 'tail'. But whether the experiment is coin-tossing or some other trial for which there are only two outcomes is irrelevant to the development of the theory. That is why so many applications of probability theory can be related to one particular type of conceptual experiment – the placing of balls in containers.

Suppose there are two balls and two containers. All possible outcomes of the experiment can be represented as in Table 3.1. If the

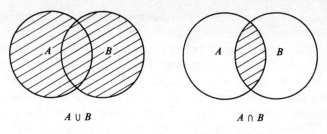

*Figure 3.2.* Venn diagrams depicting the relationships $A \cup B$ and $A \cap B$ between two sets $A$ and $B$.

balls are indistinguishable from each other there are effectively only three possible outcomes, and if the containers are indistinguishable, only two. The idea can be widened to embrace $r$ balls in $n$ containers, and applied to such diverse fields as cancer therapy, market research, accident analysis and genetics.

Modern mathematics in schools teaches the algebra of sets at an early age. However, it is repeated here to refresh the memory. $A$ is the set consisting of all the sample points defining the event $A$; $B$ is defined similarly. Then

$A'$ signifies negation, i.e. *not* the points of $A$
$A \cup B$ means either event $A$ or event $B$ or both occur
$A \cap B$ means both events $A$ and $B$ occur together
$A \cap B = \emptyset$ implies that $A$ and $B$ are mutually exclusive (they cannot occur together)

(see Fig. 3.2).

We can now associate with each event $E_i$ in the whole sample space a probability $P(E_i)$. All the events $E_i$ comprise all possible $n$ outcomes of our conceptual experiment. So if the $E_i$ are mutually exclusive we say that

$$P(E_1) + P(E_2) + \cdots + P(E_n) = 1 \qquad (3.1)$$

i.e. there is complete certainty that one of the events will occur. It follows immediately that $0 \leqslant P(E_i) \leqslant 1$.

Looking back to our set theory, we see that for any two events $E_1$ and $E_2$ the probability that either or both occur is given by

$$P(E_1 \cup E_2) = P(E_1) + P(E_2) - P(E_1 \cap E_2) \qquad (3.2)$$

If $E_1$ and $E_2$ are mutually exclusive, then $P(E_1 \cap E_2) = 0$ and

$$P(E_1 \cup E_2) = P(E_1) + P(E_2) \qquad (3.3)$$

When we toss a perfect coin we are repeating the same experiment over and over again; so too when we draw a playing card at random from a pack and replace it before drawing again. We then say that we are *sampling with replacement*. However, if each card is not replaced after it is drawn the sampling is without replacement.

With replacement sampling, from a population of $n$ elements we can draw $r$ elements in $n^r$ ways since each element can be drawn in $n$ ways. Without replacement, however, the first element can be drawn in $n$ ways, but the second in only $n - 1$ ways, the third in $n - 2$ ways, and so on. In all, there are $n(n - 1)(n - 2)\cdots(n - r + 1)$ choices. When $r = n$, the whole population is chosen each time but there are $n$ different ways of doing it. So $n$ elements can be ordered or permitted in $n!$ ways, where

$$n! = n(n - 1)(n - 2) \times \cdots \times 3 \times 2 \times 1 \qquad (3.4)$$

$n!$ is spoken as '$n$ factorial'. By Stirling's formula,

$$n! \approx (2\pi)^{1/2} n^{n+1/2} e^{-n} \qquad (3.5)$$

It is easy to see that

$$n(n - 1)(n - 2)\cdots(n - r + 1) = \frac{n!}{(n - r)!} \qquad (3.6)$$

For the famous treble chance on the football pools, the object is to pick 8 score draws from 55 matches. We have to choose a sub-population of size $r$ from a population of size $n$. In how many ways can this be done? The $r$ elements of the sub-population can be rearranged in $r!$ different ways. A sample of size $r$ can be chosen in $n!/(n - r)!$ ways without regard to order. So the number of ways in which $r$ elements can be chosen from a population of $n$ is

$$\binom{n}{r} = \frac{n!}{(n - r)!\, r!} \qquad (3.7)$$

These numbers are known as the binomial coefficients because they arise in the expansion of the expression $(p + q)^n$. They can be represented by Pascal's triangle, as in Fig. 3.3. Thus, for $n = 4$

$$\binom{4}{0} = 1, \qquad \binom{4}{1} = 4, \qquad \binom{4}{2} = 6$$

$$\binom{4}{3} = 4, \qquad \binom{4}{4} = 1 \qquad (3.8)$$

| $n = 0$ | | | | | 1 | | | | | |
|---|---|---|---|---|---|---|---|---|---|---|
| 1 | | | | 1 | | 1 | | | | |
| 2 | | | 1 | | 2 | | 1 | | | |
| 3 | | 1 | | 3 | | 3 | | 1 | | |
| 4 | 1 | | 4 | | 6 | | 4 | | 1 | |
| 5 | 1 | 5 | | 10 | | 10 | | 5 | | 1 |

etc.

*Figure 3.3.* Pascal's triangle.

In the football pools there are $\binom{55}{8} \approx 1.2$ billion ways of picking 8 matches from 55. Let us assume that the chance of each match being a score draw is $\frac{1}{4}$ (this may not be quite correct since home wins are more likely and a score draw may be more likely than an away win or a no-score draw). The chance of a selected 8 matches all being score draws is, therefore, $(\frac{1}{4})^8$, or 1 in 65 536. Thus the chance of selecting exactly 8 score draws from 55 matches with *one* solution is approximately 1 in 78 trillion (a trillion here means a million million). Of course, the whole subject of the football pools is much more complicated than this. Many entries are for multiple forecasts ('perm any 8 from 16'), and there may be a substantial winner even if there are fewer than 8 score draws.

But it is one of the peculiarities of most people's perception of probability and chance that millions of people try their luck every week. No doubt they are attracted by the prospect of an enormous prize, despite the huge odds against success.

By contrast, if you have a party of 30 people and ask them to guess the probability that two or more of them share the same birthday, the great majority of them will give wild underestimates. The correct answer is 70% – i.e. the chance of two or more people in a party of 30 having the same birthday is greater than two in three.

In making that calculation several assumptions have to be made which are not strictly accurate. For example, a year is not always 365 days long, nor are birth rates quite constant throughout the year. However, for our simple probability model we do assume these things, so that a party of 30 people may be taken to represent a random selection of birthdays. We believe that our assumptions will not materially affect the result.

The calculation is simple. For a party of $r$ people the probability that all $r$ birthdays are different is

$$p = \frac{364}{365} \times \frac{363}{365} \times \cdots \times \frac{365 - (r - 1)}{365}$$

$$p = \left(1 - \frac{1}{365}\right)\left(1 - \frac{2}{365}\right) \cdots \left(1 - \frac{r-1}{365}\right)$$

$$\approx 1 - \frac{1 + 2 + 3 + \cdots + (r-1)}{365}$$

$$= 1 - \frac{r(r-1)}{730} \qquad (3.9)$$

In this approximation, which is valid for small values of $r$, all cross-products of the fractional terms are neglected. Also, the approximation is good only for $n$ up to about 10. For larger $n$ a good approximation is then obtained by

$$\ln p = \ln\left(1 - \frac{1}{365}\right) + \ln\left(1 - \frac{2}{365}\right) + \cdots$$

$$\approx -\frac{1}{365} - \frac{2}{365} - \cdots - \frac{r-1}{365}$$

$$= -\frac{r(r-1)}{730} \qquad (3.10)$$

$$p = \exp[-r(r-1)/730)] \qquad (3.11)$$

For a comparison of these approximations, see Fig. 3.4. Suppose that we conduct a trial (e.g. the throwing of six regular dice) for which there are only two possible outcomes – success (six 6's) or failure (any other result). If the probability of success is $p$ and that of failure is $q$, then clearly

$$p + q = 1 \qquad (3.12)$$

i.e. one of these outcomes, but no other, must occur at each trial. In $n$ independent trials the probability of $r$ successes (and therefore of $n - r$ failures) is

$$B(r : n, p) = \binom{n}{r} p^r q^{n-r} \qquad (3.13)$$

because a set $r$ may be chosen from $n$ in $\binom{n}{r}$ ways.

This is the $r$th term of the binomial expansion of $(q + p)^n$, and hence we call $B(r : n, p)$ the binomial distribution of the random variable representing the number of successes in $n$ trials. We have already seen how the coefficients of the expansion of $(q + p)^n$ form the elements of Pascal's triangle (Fig. 3.3).

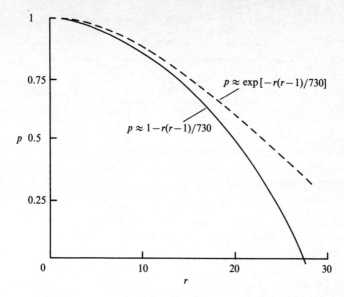

*Figure 3.4.* Approximations for the probability that $r$ people all have different birthdays. The continuous curve gives reasonable estimates for small $r$, but for larger $r$ the dashed curve is a better approximation.

In the real world there are many applications of the theory associated with these independent 'Bernoulli trials' (named after the Swiss mathematician Jacques Bernoulli). Very often $n$ is large and $p$ is small, but $\lambda = np$ is of moderate size. For these conditions, Siméon-Denis Poisson derived an approximation for $B(r: n, p)$ in the following way:

$$\begin{aligned} B(0: n, p) &= (1 - p)^n \\ &= [1 - (\lambda/n)]^n \\ &\approx e^{-\lambda} \quad \text{for large } n \end{aligned} \tag{3.14}$$

Now,

$$\begin{aligned} \frac{B(r: n, p)}{B(r - 1: n, p)} &= \frac{(n - r + 1)p}{rq}, \quad r > 0 \\ &= \frac{\lambda - (r - 1)p}{rq} \\ &\approx \lambda/r \quad \text{for large } n \end{aligned} \tag{3.15}$$

*Table 3.2.* The probability $P$ and cumulative
probability $\Sigma P_r$ that $r$ people out of 1000
share the same birthday.

| $r$ | $P(r:\lambda)$ | $\Sigma P_r$ |
|---|---|---|
| 0 | 0.0646 | 0.0646 |
| 1 | 0.1769 | 0.2415 |
| 2 | 0.2423 | 0.4838 |
| 3 | 0.2213 | 0.7051 |
| 4 | 0.1516 | 0.8567 |
| 5 | 0.0831 | 0.9398 |
| 6 | 0.0379 | 0.9777 |
| 7 | 0.0148 | 0.9925 |
| 8 | 0.0051 | 0.9976 |
| 9 | 0.0016 | 0.9992 |
| 10 | 0.0004 | 0.9996 |

It follows that

$$B(1:n,p) \approx \lambda e^{-\lambda}$$

$$B(2:n,p) \approx \frac{\lambda^2}{2!} e^{-\lambda}$$

$$\vdots$$

$$B(r:n,p) \approx \frac{\lambda^r}{r!} e^{-\lambda}$$

$$(3.16)$$

Since $\lambda = np$ we can call this

$$P(r:\lambda) = \frac{\lambda^r}{r!} e^{-\lambda} \qquad (3.17)$$

which is the *Poisson distribution* or, more correctly here, the Poisson approximation to the binomial distribution. The Poisson distribution might be called the probability distribution of 'unusual' events (since the probability of a 'success' is small).

Birthdays always make a pleasant subject – my birthday in particular. At my large birthday party on St Valentine's day there are 1000 guests. What is the probability that exactly $r$ of them also have birthdays on that day? Here $p = 1/365$, $n = 1000$ and $\lambda = 1000/365 = 2.7397$. Calculations for $r = 0, 1, \ldots, 10$ are given in Table 3.2.

We can look at Table 3.2 in a slightly different way. The probability of no one having the same birthday as me is only 0.0646. So

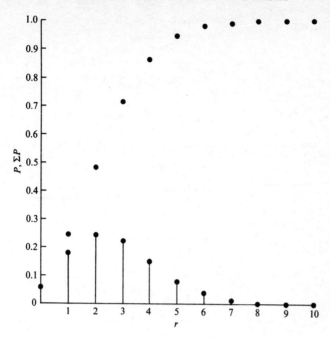

*Figure 3.5.* The cumulative distribution of probabilities given in Table 3.2.

the probability that at least one person has the same birthday is 0.9354 – rather a good chance. Looking at the right-hand column, where successive probabilities have been added together, we see that the probability of there being ten or fewer people sharing my birthday is 0.9996. So the probability of more than ten is 0.0004, a very low chance.

The figures in the right-hand column of the table are points on the *cumulative distribution*. This is the distribution of the probabilities $P(x \leqslant r: \lambda)$, plotted in Fig. 3.5. The points do not lie on continuous curves because the variable $r$ is not continuous – there cannot be $1\frac{1}{2}$ people with the same birthday.

In fact, when the variable becomes continuous in time, the Poisson distribution arises as a distribution in its own right and not just as an approximation to the binomial distribution. It has very important applications in problems connected with population growth and decline, and with waiting times in, for example, a supermarket check-out queue or a telephone system.

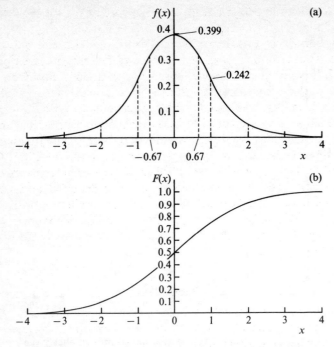

*Figure 3.6*. (a) The normal distribution curve and (b) its cumulative distribution curve. 50% of the area under the normal distribution curve lies between −0.67 and +0.67, 68.3% between −1 and +1, 95.6% between −2 and +2, and 99.7% between −3 and +3.

When the number of trials (*n*) becomes very large, these two discrete probability functions, binomial and Poisson, are approximated very closely by a famous and much used continuous function – the *normal distribution function*. This function has the equation

$$f(x) = \frac{1}{(2\pi)^{1/2}} \exp\left(-\tfrac{1}{2}x^2\right) \tag{3.18}$$

and the equivalent cumulative function is

$$F(x) = \frac{1}{(2\pi)^{1/2}} \int_{-\infty}^{x} \exp\left(-\tfrac{1}{2}y^2\right) dy \tag{3.19}$$

where $F(x)$ is a measure of the area under the curve $f(x)$ to the left of the value $x$. The two curves are shown in Fig. 3.6. With this representation a unit of $x$ is called one *standard deviation*. You may have met this term already as a measure of the 'spread' of a distribution.

The normal curve has numerous applications and occurs frequently in the natural world: it represents, for example, the height and weight distributions of human populations. An application in education is its use in the so-called 'standardization' of examination marks. The spread of marks for a large number of candidates (in a national examination, say) is assumed to follow the normal distribution, and the inherent ability of examinees is assumed to be reasonably constant over time. Differences in spread and average must, therefore, be caused by factors such as the difficulty of the particular paper or variation in the quality of teaching. To allow for this effect, all marks are adjusted or 'standardized' to fit a normal distribution with an average of 50% and a standard spread before any decisions are made on the grading marks. Needless to say, considerable controversy surrounds this practice.

## Fashion and factories

The subject of this chapter has endless fascinating ramifications. We cannot go into them all; two problems from the business world must serve to whet your appetite for more discovery.

### The buying problem

It is often possible to use the attributes of the normal distribution in problems in commerce and industry, for example the problem facing a retail buyer of merchandise in large volumes.

Suppose you are a buyer, looking to buy a new fashion item for the coming season. You know that it is highly unlikely that more than 1500 items can be sold at the suggested price, and that the most probable number is 1000. The item costs £20 and can be sold for £35. Any remaining at the end of the season will have to be cleared at £15 (i.e. a loss of £5 on each surplus item). How many should be purchased?

The intuitive answer is that the number should be such that the expected gain from selling one more item will be exactly balanced by the expected loss from failing to sell it. If $p$ is the probability of selling it, then $1 - p$ is the probability of not selling it and we have

$$15p = 5(1 - p) \qquad (3.20)$$

where £15 ($= 35 - 20$) is the profit margin and £5 ($= 20 - 15$) is the loss on clearance. From this equation, $p = 0.25$.

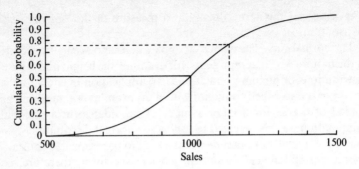

*Figure 3.7.* Cumulative probability curve for the fashion
buyer's problem.

The information given to us allows us first of all to draw a normal
cumulative distribution curve – the S-shaped curve in Fig. 3.6(b).
We assume that 'highly unlikely' means that there is only a chance
of 0.3% of sales exceeding 1500 (i.e. 1500 is 3 standard deviations
from the central value). We also assume that the 'most probable'
value (1000) occurs at the centre point of the curve – that there is a
50% chance of sales being fewer than 1000, and a 50% chance of
their being more than 1000. Because of the symmetrical shape of the
normal curve, it follows that there is a 99.7% chance of sales exceeding
500. So the cumulative probability curve for sales looks like Fig. 3.7.

A point on this curve gives the probability that sales will *not*
exceed a given level. We want the point on the curve where the
probability is 0.25 that sales *will* exceed the corresponding level, i.e.
where the probability is 0.75 that sales will *not* exceed this level. The
dashed lines in the diagram link the probability 0.75, via the curve,
to the corresponding sales level. This is approximately 1170, and is
the number of items you should buy.

Remember that there is no certainty in this answer: no one knows
how many will in fact be sold. It could be only 100, in which case the
financial outcome would be, from equation (3.20),

$$£(100 \times 15) - (1070 \times 5) = £1500 - 5350$$

i.e. a loss of £3850. Or you could sell 1170, making a profit of

$$£1170 \times 15 = £17\,550$$

There is a very big difference between these two figures: one is a
dismal failure, the other a resounding success. A buyer's life is never
easy but the above is a rational way to make a decision in the face
of uncertainty.

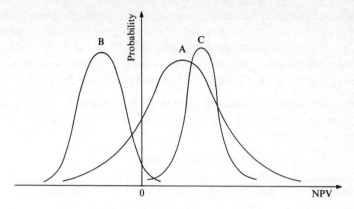

*Figure 3.8.* Net present value (NPV) probability curves for three new factory projects, A, B and C.

### The investment problem

Now we move from retailing to manufacturing. When a company is considering building a new factory the board of directors will probably ask for forecasts of the profits likely to arise from different factory capacities. Since future demand for the factory's products, and the building costs and time to completion of the new facility, cannot be known with certainty, neither can the expected pay-off from the investment.

The cash-flows (flows out for capital expenditure and costs; flows in for sales income) for any particular investment can be discounted back to the present by a discount factor representing the cost of borrowing money (interest rate) plus a profit margin to offset the risk of the investment. The sum of the discounted cash-flows is known as the net present value (NPV) of the investment. If the variables in the project are subject to probability distributions, then so too will be the NPV.

Thus the directors may be faced with, say, a choice between three factories A, B and C of different sizes, or possibly of the same size but in different locations. The NPV probability of these may differ widely; they might look like Fig. 3.8. Which project should be chosen by the board? Project A shows a significant probability of making a loss, but also a significant probability of substantial profits. Project B appears almost certain to make a loss. Project C is almost certain to achieve a modest profit.

It is fairly easy to reject B, but a choice between A and C will depend, among other things, on the 'risk aversion' of the directors (i.e. do they like to play safe?) and also on the pay-off relative to the amount of money to be invested. If the investment is very large in relation to the size of the company, then C would be a sensible choice. If, however, the investment is only a modest sum, then it could well prove worth while to take a chance with A.

Decisions cannot be removed from those responsible by these kinds of analysis. There are rarely easy answers in this difficult and uncertain world.

> We do not what we ought;
> What we ought not, we do;
> And lean upon the thought
> That chance will bring us through.

So often it does not!

### Further reading

B. Edwards, *The Readable Maths and Statistics Book* (George Allen & Unwin, 1980).

G. Humphreys, *Turning Uncertainty to Advantage* (McGraw-Hill, 1979).

M. J. Moroney, *Facts from Figures* (Penguin, 1965).

P. Sprent, *Taking Risks* (Penguin, 1988).

# 4

INTEGRATING: MIXING PAINT,
NEWSPAPER PICTURES AND
CAR EXHAUSTS

JULIAN HUNT AND DAVID STURGE

### Some new questions to consider

There is a nice story about Dido, the legendary queen who had a
tragic love affair with Aeneas as he journeyed back from Troy. When
she first arrived as a refugee in North Africa, she was offered as large
an area on the coast as could be enclosed by the hide of a cow. She
took the hide and cut it into long, thin strips, which she sewed
together and then laid out on the ground. What shape of curve
should she have chosen so as to enclose the maximum area $A$ within
the perimeter of the hide, of length $L$ (see Fig. 4.1)? According to the
story the area she managed to enclose was large enough for her to
found the ancient city of Carthage!

Mathematically, Queen Dido had to choose a function $f(x)$ to
meet a particular requirement. There are many practical examples of
this kind of mathematical problem. It differs from problems (such as
those we met in Chapter 1) in which we have to *analyse* functions
presented to us by particular situations or by scientific laws.

The theme of this chapter is the choosing and approximating of
functions that we require in order to define curves or surfaces or
pictures. Faced with the problem of creating (or synthesizing) functions as well as analysing them, we need to develop some new ideas
and techniques. We find that integral calculus is the connecting
theme in both the choosing and the approximating. Three areas of
application of this field of mathematics are the design of processes
and of products, and communications.

An example of a *process* problem is the design of machines for
mixing different colours of paint or for mixing food. How would you
decide how beaters in a food mixer should move – should the beaters
make the food just go round in circles, or move in figures-of-eight,

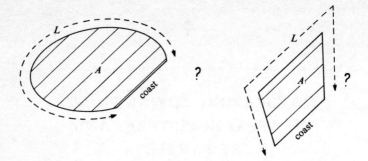

*Figure 4.1.* Dido's problem: what shape of curve will give the greatest area *A* enclosed within a perimeter of length *L* and the coastline?

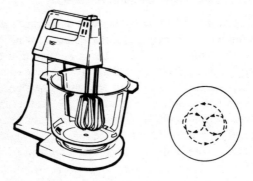

*Figure 4.2.* The food mixer problem: what sort of motion should the beaters impart to the food – circular, maybe, or figure-of-eight?

or what (see Fig. 4.2)? There are two mathematical problems to be solved here: to analyse the mixing process, and to choose the curves or paths of the mixer that will bring about the most effective mixing; of course it is also necessary to consider whether a practical mixer can be designed to meet this objective.

The designer of a new plastic jug and an architect designing the roof of a new sports stadium are both concerned with *products*. With the new materials and new manufacturing techniques that are now available they could choose almost any shape. But they have to bear in mind the volume of the contained space, the surface area of the container, the strength of the container and many other factors. After choosing the shape it would be necessary to *define* the shape mathematically and then calculate all these properties. With new computers and programs, all these steps are becoming very much easier for industrial designers and engineers.

Mathematics has also played an essential role in the development of new *communications*, the most obvious technological revolution of recent times. Pictures on TV and in newspapers, transatlantic telephone calls, and sending messages through the mud of an oil well – all these are different examples of modern communications, but similar mathematics lies behind each one.

The black and white newspaper picture in Fig. 4.3 is made up of thousands of tiny dots which are either black or white. Together, the human eye and brain find that this bears a close resemblance to an

*Figure 4.3.* (a) A photograph and (b) its reproduction as a pattern of dots in a newspaper picture. The dots are clearly visible in the enlargement (c). Photograph: Frank Baron/*Guardian*

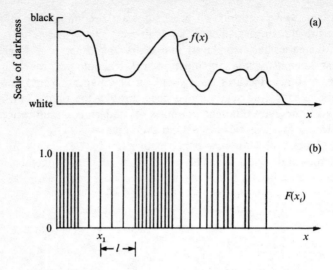

*Figure 4.4.* The function $f(x)$ in (a) is 'equivalent' to the
distribution of spikes in (b). This is the mathematical
equivalent of representing the photograph in Fig. 4.3(a) as
the pattern of dots in Fig. 4.3(b).

actual photograph made up of continuous shades of grey between
white and black. But in fact the actual photograph, from which the
newspaper picture comes, is also a speckled surface of black and
white grains of chemical, but on a much finer scale. (We are in a
'grey' area of imprecise words and ideas!) So what is the *mathematical* requirement for a satisfactory picture, as perceived by the eye,
and how many dots are really needed? It is a question of representing
a continuous function $f(x)$ (where $f$ lies between 0 and 1, and $x$
between 0 and 10) by another function $F(x_i)$, where $F$ is either 0 or
1 at a large number (say 1000) of distinct points, $x_1, x_2, x_3, \ldots$ (the
$x_i$), lying between 1 and 10 (see Fig. 4.4). In Chapter 1 (Fig. 1.6(a))
we approximated a curve by a large number of straight lines. We
obviously need to take a different approach to this problem.

We must begin the mathematics in this chapter with a brief
introduction to integrals and how they relate to *differentiating*
functions. We'll look at some interesting examples along the way.

### Areas and integrals

If, like Dido, you had to choose a particular curve so as to enclose
the largest area, $A_{max}$, you would first have to be able to calculate the

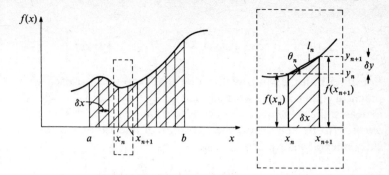

*Figure 4.5.* Evaluating the area under a curve. The area
between $x = a$ and $x = b$ is divided into strips, one
of which, between $x = x_n$ and $x = x_{n+1}$, is shown in the
enlargement.

area $A$ under *any* curve $f(x)$. (We'll discuss how to choose $f(x)$ so as
to maximize $A$ later.)

So first we look at how to calculate the area between the $x$ axis and
the part of a general curve $f(x)$ lying between the lines $x = a$ and
$y = b$ (Fig. 4.5). Since this area depends on the values of $a$ and $b$,
we express $A$ as $A = I(a, b)$, because it is a function of $a$ and $b$.

Just as we did when calculating the slope of a curve and differentiating it, when we calculate the area $A$, we also begin by
dividing the interval $(a, b)$ into $N$ thin strips of equal width, $\delta x$. We
take one strip, between $x_n$ and $x_{n+1}$, say. Now, the bit of the curve
$f(x)$ across the top of the strip is *almost* a straight line, so the strip
has about the same area as the shaded trapezium (a quadrilateral
with parallel sides), which is $\frac{1}{2}\delta x[f(x_n) + f(x_{n+1})]$, where $\delta x =
x_{n+1} - x_n$. There is a small error in this expression only if there is
some curvature in the function $f(x)$, i.e. if $d^2f/dx^2$ is non-zero. But the
narrower the strip, the smaller the error.

By adding together the areas of all the strips between $x = a$ and
$x = b$, we obtain an approximation for the whole area which we
denote by

$$I_N(a, b) = \sum_{n=0}^{N} \tfrac{1}{2}\delta x[f(x_n) + f(x_{n+1})] \qquad (4.1)$$

(where the sign $\Sigma$ means summing all the values between $n = 0$ and
$n = N$.) As all the strips have the same width, we can simplify this

expression to

$$I_N(a,b) = \delta x\left[\tfrac{1}{2}f(x_0) + \sum_{n=1}^{N-1} f(x_n) + \tfrac{1}{2}f(x_N)\right] \qquad (4.2)$$

For almost all the kinds of function that crop up in science or in practical applications, we find that as the *number* of strips, $N$, increases, the value of $I_N(a,b)$ tends to a *limit*. We call this limit the *integral* $I(a,b)$, and denote the operation of integrating $f(x)$ by the curly integral sign, $\int$. So:

$$I(a,b) = \lim_{N\to\infty} I_N(a,b) = \int_a^b f(x)\,dx \qquad (4.3)$$

The $b$ and $a$ above and below the integral sign indicate that $f(x)$ is to be integrated between $a$ and $b$.

For most functions or curves the area or integral has to be computed by an approximate method, similar to the strip method. Usually this tends to an accurate estimate of $I$. For example, if we compute the area under the curve $e^x$ from $x = 0$ to $x = 1$ by summing the area of strips, we find that if we take only one strip ($N = 1$) then $I_1 = 1.8591\ldots$; if we take ten strips ($N = 10$) then $I_{10} = 1.7187\ldots$. For $N \to \infty$, we can show that $I = 1.7183\ldots$

Note that all the approximate values are *above* the actual value because the trapeziums are above the curve. Can you think which kinds of curve or function would prove much more difficult to integrate? Draw a few curves and then try drawing the relevant strips.

### *An integral is more general than an area*

Instead of talking about *areas* we shall now use the term *integrals*, because integrals are used in connection with all sorts of functions – not just those representing curves on a surface.

In practice, the process of finding the integral of a function is the inverse of differentiation. (This is proved in standard A-level textbooks.) For example, $\int x^2\,dx$ is equal to $\tfrac{1}{3}x^3 + k$ because if we differentiate this we get $x^2$. This is known as an *indefinite* integral because we cannot determine the constant $k$ without having some information about the problem we are trying to solve.

For a *definite* integral (often an area), we would have

$$\int_a^b x^2\,dx = \left[\frac{x^3}{3}\right]_a^b = \frac{b^3}{3} - \frac{a^3}{3} \qquad (4.4)$$

This gives the value of the area under the graph between $x = a$ and $x = b$, and this area may in its turn represent, for example, a distance or a total flow rate. In Chapter 1 we met a problem where a particle was projected vertically and its velocity was found to be $v(t) = 20 - 10t$. If we now want to find how far it will travel in the first second, we can write $v = dy/dt = 20 - 10t$, so that

$$y(1) = \int_0^1 (20 - 10t)\,dt = \left[20t - 5t^2\right]_0^1 = 15 \quad (4.5)$$

where our units are metres.

In the next second, the projectile will rise by 5 m, since

$$\int_1^2 (20 - 10t)\,dt = \left[20t - 5t^2\right]_1^2 = 20 - 15 = 5 \quad (4.6)$$

We can see that for some mathematical functions integration gives another mathematical function with known properties (see also equations (1.26) to (1.31) in Chapter 1). Tables of integrals are given in standard texts.

However, there are many functions that do *not* have integrals equal to other known functions (e.g. $\int_0^x \sqrt{(\sin u)}\,du$). For these we have to *compute* the definite integral by using approximate methods of finding the area under the curve, such as equation (4.2). So far in this chapter we have looked at standard mathematics, which you may have been surprised to find in a book on new applications. However, as we shall now see, there are many new ways in which integration is used in problem-solving.

## The length of a line and choosing the right curve

### *The length expressed as an integral*

In Dido's problem the length of the line $L$ is known, but the function $f(x)$ that determines the shape is not. Let us tackle the problem by assuming that we know the answer. We then ask another question: How do we calculate the length of the curve $y = f(x)$ between $x = a$ and $x = b$? We can obtain different approximate answers to this question by considering, as we did for Fig. 4.5, a series of $N$ separate values of $x$ between $a$ and $b$, namely $0, x_1, \ldots, x_n, \ldots, x_N$. Then we approximate the perimeter of the curve by summing the lengths of the $N$ straight lines between the points on the curve next to each other, e.g. between $y_n = f(x_n)$ and $y_{n+1} = f(x_{n+1})$.

By Pythagoras' theorem, the length $l_n$ of the straight-line element between the points $(x_n, y_n)$ and $(x_{n+1}, y_{n+1})$ is

$$l_n = \sqrt{[(x_{n+1} - x_n)^2 + (y_{n+1} - y_n)^2]} \qquad (4.7)$$

Note that, if the line is at an angle $\theta_n$ to the axis (see Fig. 4.5), then by trigonometry

$$l_n = \frac{x_{n+1} - x_n}{\cos \theta_n} \qquad (4.8)$$

The sum of these lengths can be written as

$$L_N = \sum_{n=0}^{N-1} l_n \qquad (4.9)$$

You can see by eye that taking more and more elements between $x = a$ and $x = b$ will yield sets of straight lines approximating ever more closely to the curve. As $N \to \infty$, the line length tends to the limiting value

$$L = \lim_{N \to \infty} L_N = \lim_{N \to \infty} \sum_0^N \sqrt{\left[1 + \left(\frac{\delta y}{\delta x}\right)^2\right]} \, \delta x$$

But as $\delta y \to (\mathrm{d}y/\mathrm{d}x)\,\delta x$, and as $\delta x \to 0$, the sum becomes an integral; so we find that the *length* of a line is also an integral:

$$L = \int_a^b \sqrt{\left[1 + \left(\frac{\mathrm{d}y}{\mathrm{d}x}\right)^2\right]} \mathrm{d}x \qquad (4.10)$$

In Chapter 1 we saw that $\mathrm{d}y/\mathrm{d}x = \tan \theta$. For some calculations it is easier to express this integral in terms of $\theta(x)$ using equation (4.8):

$$L = \int_a^b \frac{1}{\cos \theta(x)} \, \mathrm{d}x \qquad (4.11)$$

The integral (4.10) is generally difficult to calculate exactly. But if the curve is a straight line between $y = y_0$ at $x = a$ and $y = y_N$ at $x = b$, the value of $\mathrm{d}y/\mathrm{d}x$ is just the constant slope $(y_N - y_0)/(b - a)$. You should substitute this value of $\mathrm{d}y/\mathrm{d}x$ into equation (4.10) and check the answer with Pythogoras!

Another perimeter is the circumference of a circle. This is a little more difficult to calculate; we use equation (4.11) and some trigonometry. Let the circle be centred at $x = 0$, $y = 0$ (see Fig. 4.6), and let its radius be $a$. Since all the points $(x, y)$ on its perimeter are at the same distance from its centre, then by Pythogoras'

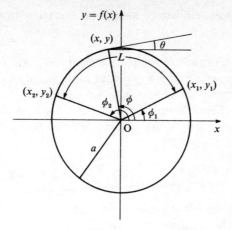

*Figure 4.6.* Calculating the distance round a circle from $(x_1, y_1)$ to $(x_2, y_2)$ in terms of the tangent angle, $\theta$.

theorem

$$x^2 + y^2 = a^2 \tag{4.12}$$

So the function $f(x)$ defining the curve is

$$y = f(x) = \sqrt{(a^2 - x^2)} \tag{4.13}$$

The values of $y$ and $x$ can be expressed in terms of the angle $\phi$ between the radius and the $x$ axis as

$$y = a \sin \phi \quad \text{and} \quad x = a \cos \phi \tag{4.14}$$

Differentiating these expressions shows how $y$ and $x$ vary with $\phi$:

$$\frac{dy}{d\phi} = a \cos \phi \quad \text{and} \quad \frac{dx}{d\phi} = -a \sin \phi \tag{4.15}$$

and so

$$dy = a \cos \phi \, d\phi \quad \text{and} \quad dx = -a \sin \phi \, d\phi \tag{4.16}$$

(assuming that we can justify splitting the derivative and that $dx$, $dy$ and $d\phi$ have meaning). Dividing equations (4.16) one into the other shows how $y$ varies with $x$:

$$\frac{dy}{dx} = -\frac{1}{\tan \phi} \tag{4.17}$$

But $dy/dx$ is equal to $\tan \theta$ (see Fig. 4.6), so by trigonometry the angles $\theta$ and $\phi$ are related:

$$\theta = \phi - 90° \tag{4.18}$$

Therefore the length of the perimeter of the circle between the points where $x = x_1$ and $x = x_2$ is the same as the length between angles (now defined in radians) $\phi_1$ and $\phi_2$, where $x_1 = a\cos\phi_1$ and $x_2 = a\cos\phi_2$. This is called changing the variable in an integral – a useful trick for simplifying the calculations. So we can change equation (4.11) to

$$L = \int_{\phi_1}^{\phi_2} \frac{1}{\sin\phi} (-a\sin\phi)\,d\phi = a(\phi_2 - \phi_1) \tag{4.19}$$

If $\phi_2$ is $2\pi$ radians (360°) greater than $\phi_1$, it is at the same point on the circle. So the perimeter of the whole circle is $2\pi a$.

*Fencing land or losing less heat by maximizing areas*

If you are enclosing the maximum area of land for a given length of fence (or cowhide in Dido's case), how should you set about erecting the fence? A similar problem is determining the shape of a pipe or duct carrying warm air so that it loses the least amount of heat from its surface for a given flow along the pipe. The mathematical question is, how do we maximize the area within a curve $y = f(x)$ for a given length of curve $L$?

Let us assume that the ends of the curve are to lie on the $x$ axis, $y = 0$. We have to find the maximum value of the integral

$$A = \int_a^b f(x)\,dx \tag{4.20}$$

for a given length of the curve

$$L = \int_a^b \sqrt{\left[1 + \left(\frac{df}{dx}\right)^2\right]}\,dx \tag{4.21}$$

where $f(x) = 0$ at $x = a$, $x = b$ $(b > a)$ and $L > b - a$. The values of $a$ and $b$, and the function, have to be found.

Doing mathematics is often like doing an experiment. When you have a new or difficult problem, you should try different approaches using whatever techniques you already know. So if you (say as a sixth-former) were faced with Dido's problem, you might start by drawing polygons, or other shapes with straight edges, based on the end points. For each figure you could calculate the area $A$ and perimeter $L$ without using integrals.

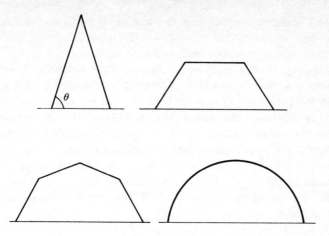

*Figure 4.7.* Dido's problem again: how to increase the area
enclosed by a given perimeter.

The simplest figure would be an isosceles triangle with the sides at
an angle $\theta$ to the $x$ axis (see Fig. 4.7). Given the total length $L$ of the
two equal sides, its area would be

$$A = \tfrac{1}{4}L^2 \cos \theta \sin \theta \qquad (4.22)$$

You could find the value of $\theta$ corresponding to the largest value of
$A$, either by using calculus (by differentiating equation (4.22) with
respect to $\theta$) or by drawing a graph of $A$ against $\theta$. You would find
this value of $\theta$ to be 45°, and the maximum value of $A$, $A_{max}$, to be
$\tfrac{1}{8}L^2$. (So the shape would be half a square.) Note that $A$ (which
would be measured in, say, square metres) must be proportional to
the square of the length (measured in, say, metres). Therefore the
ratio $A_{max}/L^2 = 0.125$ for any *size* of triangle would be independent
of the units used.

If you next took polygons on our 'coast' with three equal sides and
then with four equal sides, you would find that $A_{max}/L^2$ would
increase, to 0.144 and 0.151 respectively. So it appears that by
increasing the number of sides we are continually increasing $A_{max}$ for
given $L$. You can probably guess what shape will give the largest
area. It is, of course, the *circle* (as Dido knew, apparently). For the
semicircle of radius $a$, the ratio is

$$A_{max}/L^2 = \tfrac{1}{2}\pi^2/(\pi a)^2 = 1/2\pi = 0.159 \qquad (4.23)$$

Therefore the increase in area in going from four sides to an infinite
number of sides would be less than 10%! There are many other

applied mathematics problems in which we want to find the maximum of a quantity that depends on the integral of an unknown function. Making an intelligent guess for the function – in this case the curve $f(x)$ – often gives an estimate correct to within 10% for the maximum values (in this case the quantity is the area). Even if we guess a *function* which is more than 10% in error (i.e. the wrong curve in this case), the error in the *integral* will be less than 10%. Frequencies of vibration of systems ranging from large structures to molecules can be calculated with surprising accuracy by calculating the maximum energy associated with certain assumed (or 'guessed') patterns or forms of vibration. There is no need to compute the precise details of the vibration to obtain useful estimates of the frequencies.

## *Mixing paint or food by maximizing lengths of curves*

Now we turn this mathematics round and apply it to mixing paint or mixing food. This is one of many applications where we are interested in how lengths of curves or surfaces can be *maximized*. (A nice thing about applied mathematics is the way the applications can change suddenly when one asks a slightly different question about the same mathematical analysis.)

When two ingredients, which we shall label F and B, are thoroughly mixed, any sample of the mixture, of whatever size, has exactly the same proportion of the ingredients. (When you are making scones by mixing flour (F) and butter (B), you should not end up with any lumps of butter in the mixture – it should be very smooth.) So in the *process* of mixing, the ingredients F and B must be spread throughout the mixing volume (see Fig. 4.8).

In mathematical terms we have to be a little more precise about these ideas. We define mixing as being 'completed' down to a scale, $l_m$, which is much less than the size of the mixing vessel. This means that in any 'completely' mixed sample larger than $l_m$ the proportions of F and B are the same (i.e. F and B are well mixed), but in a sample smaller than $l_m$ there may be differences in the proportions (i.e. F and B are not well mixed on this small scale).

When we start mixing, F and B are separated by a surface, S. As the mixing develops, this surface becomes highly distorted and winds throughout the volume. We divide the volume into boxes of scale $l_m$. If the area across the mixing vessel is $A$, the number of boxes is proportional to $A/l_m^2$. When mixing is completed down to the scale $l_m$, the surface (or *interface*) between F and B *must* cross every box

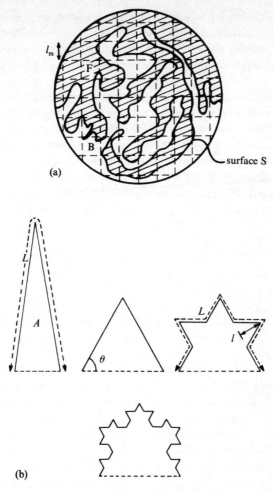

*Figure 4.8.* (a) A contorted surface S separating F and B,
two ingredients undergoing mixing. The length $l_m$ is the scale
of the mixing. (b) A simple model for studying how the
mixing process is improved by making the surface S more
'wiggly'.

of side $l_m$. So the length $L$ of the interface must be equal to or greater
than the number of boxes times the length of the box:

$$L \geqslant (A/l_m^2)l_m = A/l_m \qquad (4.24)$$

Therefore a good way of assessing the effectiveness of a mixing
process – of finding a 'quantitative measure' – is to look at the quantity

$l_m^{-1}$, or equivalently the ratio $L/A$. The smaller $l_m$ (or the larger $l_m^{-1}$ and $L/A$), the more effective the mixing. Note that the ratio $L/A$ increases because the length of the interface $L$ increases; in fact the area $A$ remains constant (in this simplified model of two-dimensional mixing).

In most processes involving liquids and gases, the actual molecules of the ingredients mingle with each other across the interface S. The interface then becomes a 'fuzzy' region of F and B, so that, in mixing paint or mixing milk and tea, the mixing is eventually completed down to the scale of the molecules. Most chemical reactions can proceed efficiently only if mixing is complete down to this scale. So in chemical processes, mixing on the molecular scale is essential whatever the size of the vessel – whether it is a test-tube, which you shake for mixing, or a huge tank for making plastics, where the mixing is done by large rotors.

Suppose we take the simplest, *ideal* mixing process in which the interface S between F and B, shown in Fig. 4.8(a), is modelled by assuming it to be distorted into a series of triangles, as shown in Fig. 4.8(b).

In equation (4.22), we found the area $A$ of the triangular shape for a given length $L$. If we now ask what is the *maximum total length* of the two sides $L_{max}$ (say in metres) for given area $A$ (say in square metres), we shall need to find the shape that gives the maximum value of $L/\sqrt{A}$ (to have an answer that is independent of the units or the size of the shape). So, as the angle $\theta$ changes, $A$ changes if $L$ is constant. Therefore $L$ must change if the total area $A$ is constant. (In Fig. 4.8(b) neither $A$ nor $L$ is constant, but the shapes illustrate the point being made here.)

If the shape is a triangle, then the formula (4.22) gives the ratio

$$\frac{L}{\sqrt{A}} = \frac{2}{\cos\theta\sin\theta} \tag{4.25}$$

Since $\cos\theta = 0$ when $\theta = 90°$ and $\sin\theta = 0$ when $\theta = 0°$, $L/\sqrt{A}$ tends to $\infty$ as $\theta$ tends to 0 or 90°.

In a given vessel with diameter $L_v$, the sum of the lengths of the two sides of the triangle cannot be greater than $2L_v$. So in mixing food (or anything else), you increase the length $L$ of the interface in two ways: first by wrapping the interface into curved shapes (e.g. spirals) within the vessel, and secondly (if the mixture is not too sticky) by creating lots of irregular 'wiggles' on the interface between the ingredients, with a small length scale $l$ (the subscript m is omitted, for simplicity).

Suppose that, on two of the sides of a 60° triangle, of combined length $L_0$, you create two others with side one-third of the original triangle (see Fig. 4.8(b)). Then the length of the interface, $L$, has increased to $(4/3)L_0$. So has the area, but only to 11/9 of its previous value: the increase in area is relatively less than in length. In fact $L/\sqrt{A}$, 3.04 for the simple triangle, has increased to 3.67. If this is repeated, adding eight new little triangles, as in Fig. 4.8(b), $L$ increases to 4/3 of the previous value again and $A$ increases by 4%; the ratio $L/\sqrt{A}$ is increased to 4.79. Repeating this process $n$ times leads to $L/\sqrt{A}$ increasing in proportion to $L_0(4/3)^n$, since $A$ tends eventually to a fixed limit. At the same time $l$ decreases to $(1/3)^n$ of $L_0$.

This shows that, as the scale of the 'wiggles' decreases, the overall length of the surface $L$ increases. But how are these two lengths related? Taking logarithms, $\ln(L/L_0)$ is proportional to $n\ln(4/3)$, and $\ln(l/L_0)$ is proportional to $n\ln(1/3)$. Eliminating $n$, we find that

$$\frac{\ln(L/L_0)}{\ln(l/L_0)} = \frac{\ln(4/3)}{\ln(1/3)} = -r, \text{ say} \qquad (4.26)$$

Therefore $L$ and $l$ are related by

$$\ln(L/L_0) = \ln[(l/L_0)^{-r}] \qquad (4.27)$$

or

$$(L/L_0) = (l/L_0)^{-r} \qquad (4.28)$$

and

$$r = -\ln(4/3)/\ln(1/3) = 0.261 \qquad (4.29)$$

For a straight line, $r = 1$.

This idealized example shows how, by reducing the scale of the wiggles by a factor 10, say (which could be done by vigorous stirring), the total length $L$ of the interface is increased by a factor of 18.

There are many industrial and natural processes where efficient mixing between different ingredients or substances in liquids and gases is made possible by small-scale chaotic motion, which we call *turbulence*. In the chemical and food-processing industries mixing is followed by chemical reactions which convert the raw materials into the desired products. The mixing processes are sometimes estimated by calculating how the interface between the ingredients evolves and how chemical reactions occur across the interface.

The areas or lengths of the wiggly surface of lines $L$ always increase as the lengths of the wiggles are decreased. This is because the expression (4.10) for $L$ involving an integral of $(dy/dx)^2$ shows that $L$ increases as $dy/dx$ increases and, for a *smooth curve*, $dy/dx$ increases as $l$ decreases. As we discussed in Chapter 1 (see Fig. 1.16), there are many types of curve where the scale of wiggles depends on how closely the curve is inspected. In the 1920s, in one of the first studies of this subject, Lewis Fry Richardson found that the length of the frontiers between countries kept increasing as he measured on a smaller and smaller scale. Benoît Mandelbrot revived Richardson's work in the 1960s, asking 'How long is the coast of Britain?' (see Chapter 6).

### Exhaust pipes and plastic jugs – designing curves and surfaces with computers

#### Computer-aided design

Imagine you are a designer of sports cars. You have 'sketched' and worked out a beautiful aerodynamic shape, with curves and angles never seen before. So you excitedly show this to the production engineers who have got to make it. What do you think they would say? 'You must be joking! How are we going to make that shape?' Or, 'Fine, just run it through the CAD [computer-aided design] system and I'll check with the structures and production staff that the metal sheet at those angles will be strong enough.' Nowadays something like the second reply is what the designer expects, because much of the 'sketching' is done on a computer screen with a light-pen.

Once a sketch has been made, the computer will calculate and then draw other views of the car. Where it does not have enough information (because the sketch does not include all the faces of the car) it makes a guess. Then, just as a sculptor walks around his sculpture in breaks between the moulding or chipping, the computer designer – without having to get up – can look at different perspectives.

Before the use of CAD systems, mock-up models (of clay and wood) used to be built, and the coordinates (i.e. the precise location $x, y, z$) of points on the surface of the design had to be measured from the model or from drawings. This could lead to variations in the product, because engineering drawings of views from the three directions do not precisely define complicated surfaces. With CAD, the coordinates are given automatically. Other

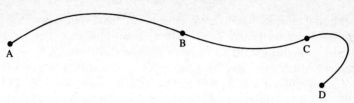

*Figure 4.9.* The kind of shape designers need to define
mathematically – a car exhaust.

quantities too can easily be calculated, such as the length of a line
along a curve, the area of a curved surface, the cross-sectional area
or the volume within a closed surface. In more recent systems, the
coordinates and other details of the design are fed into another
program for computing structural strength.

This revolution in design has come about with the help of
engineers, computer scientists and mathematicians. Their particular
contribution has been to find simple and accurate mathematical
expressions to *describe* the curves and surfaces that arise in engineering designs. Part of the mathematical analysis is now to *choose* the
best functions to meet special needs (such as reducing heat transfer
or increasing strength) – the topic of the previous section. Now we
concentrate on the description of curves and, to a lesser extent,
surfaces.

### Curves

Figure 4.9 shows the curly shape of the exhaust pipe of a car. It is
normally made and sold as one piece. In manufacturing such a
complicated shape, it is important to have an equation that expresses
the shape. Even though computers are used, we need to look for an
equation that not only represents the curve we are interested in, but
also has coefficients that are easy to *understand*. (Remember that
computer programs have to be understood too!)

As an example of an unsuitable equation, we need only look at an
ordinary polynomial,

$$y = a_0 + a_1 x + a_2 x + \cdots + a_n x^n \qquad (4.30)$$

It fails to satisfy either of our two requirements, because for each $x$
there is only one $y$ – so we cannot make closed loops – and it is
almost impossible to understand the effect of changing one of the
coefficients $a_i$ $(0 \leqslant i \leqslant n)$.

Splitting our required curve into *pieces*, and then applying a separate equation to each piece, makes the problem rather easier. It is more straightforward to imagine finding a curve with a simple equation which looks like the portion AB in Fig. 4.9, another curve with an equation like BC, and a third with an equation like CD. This is better than trying to find a curve equation that 'looks like' the whole of ABCD in one go. The question is, what does 'look like' mean?

Obviously we must end up with 'reasonably' smooth curves – usually the function and the first differential are continuous, but the second differential (or curvature) may change across the join. To avoid any difficulties with this discontinuity, smaller lengths of the curve are approximated where the curvature changes rapidly (as with CD). In this example, we might get away with making each portion a circular arc. In fact, many commercial CAD systems represent all curves by combinations of straight-line and circular segments. Some more recent systems are better than the line-arc scheme, and overcome the restriction of the ordinary polynomial being single-valued. To do this they use two simultaneous equations for $x$ and $y$ in terms of a *parametric variable*, $t$:

$$\begin{aligned} x &= a_0 + a_1 t + a_2 t + a_3 t^3 \\ y &= b_0 + b_1 t + b_2 t + b_3 t^3 \end{aligned} \tag{4.31}$$

This is much more flexible than expressing $y$ as a function of $x$. A similar equation for $z$ is used when working in three dimensions.

The *parametric cubic* is a useful curve for piecewise curve design because it can represent inflected curves (ones whose curvature changes sign) as well as uniformly curved segments. In the form of equations (4.31), it is still difficult to see what the coefficients mean. However, Bézier had the simple but clever idea of rearranging the coefficients as follows:

$$\begin{aligned} x &= c_0(1 - t)^3 + 3c_1 t(1 - t)^2 + 3c_2 t^2(1 - t) + c_3 t^3 \\ y &= d_0(1 - t)^3 + 3d_1 t(1 - t)^2 + 3d_2 t^2(1 - t) + d_3 t^3 \end{aligned} \tag{4.32}$$

You can see that this is still a cubic, but now the coefficients have a geometrical meaning, shown in Fig. 4.10.

Let $t = 0$ in equations (4.32). Then $(x, y) = (c_0, d_0)$ at the point on the curve where $t = 0$. By differentiating with respect to $t$, and calculating $dy/dx$ at $t = 0$, we can show that $(c_1, d_1)$ lies on the tangent to the curve at $t = 0$. Similarly, we find that $(c_2, d_2)$ lies on the tangent to the curve at $t = 1$, as does the point $(c_3, d_3)$. Note that

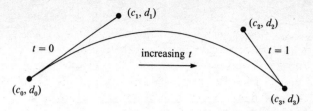

*Figure 4.10.* Defining an element of a curve as a Bézier cubic polynomial with easy-to-understand coefficients.

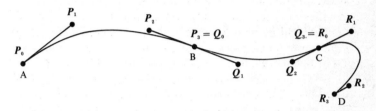

*Figure 4.11.* The control points for the car exhaust curve (Fig. 4.10) after division into three cubic polynomial elements.

while two of these points do not actually lie on the curve, their positions are both meaningful.

We now introduce some vector notation. We write $r = (x, y)$ as a point on the curve, and we let $P_i = (c_i, d_i)$ be the 'control points' (for $i = 0, 1, 2, 3$) of the curve. We then write

$$r = P_0 t^3 + 3P_1(1 - t)^2 + 3P_2 t^2(1 - t) + P_3 t^3 \quad (4.33)$$

This 'vector equation' means that there are two or three separate equations to be evaluated at the same time, one for each of the $x$, $y$ and (maybe) $z$ components. For the next sections of the curve the control points are at $Q_i$, $R_i$, etc.

We can now add together the pieces of the curve of the car exhaust – see Fig. 4.11. Place $P_0$ at A, $P_3$ and $Q_0$ at B, $Q_3$ and $R_0$ at C, and so on, to make the ends of the pieces match up in position. Place $P_2$, $P_3 = Q_0$ and $Q_1$ on one straight line, and $Q_2$, $Q_3 = R_0$ and $R_1$ on another straight line to ensure that the end tangents of the pieces line up. We then apply the *Bézier parametric cubic* (4.33), with $t$ going from 0 to 1, and the $P$ coefficients to get the piece from A to B, the same equation with $t$ going from 0 to 1 and the $Q$ coefficients to get the piece going from B to C, and the same equation with $t$

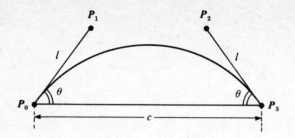

*Figure 4.12*. The control points for a circular arc element.

going from 0 to 1 and the **R** coefficients to get the piece going from C to D.

Here we have a method of representing any shape of curve we like by using one simple equation repeatedly, with different coefficients. Quite often we want to represent circular arcs in a curve. Fortunately the Bézier parametric cubic can represent a circular arc to 'engineering' accuracy, as long as we put the control points in the right place.

If $c$ is the chord length $|P_3 - P_0|$ and $\theta$ is the angle between the tangents and the chord (see Fig. 4.12), then the rule is to place $P_1$ and $P_2$ so that

$$l = |P_1 - P_0| = |P_3 - P_2| = \frac{2c}{3(1 + \cos\theta)} \quad (4.34)$$

For $\theta < 45°$ (a quarter-circle), the parametric cubic then lies within $0.0005c$ of the circular arc with the same end tangents.

The area under the curve, in the $x$–$y$ plane, can be obtained by calculating the area under a Bézier parametric segment between $x = c_0$ and $x = c_3$. This is given by

$$A = \int_{c_1}^{c_3} y \, dx = \int_{t=0}^{t=1} y(t) \frac{dx}{dt} \, dt \quad (4.35)$$

So

$$A = \int_0^1 \left\{ [d_0(1 - t)^3 + 3d_1 t(1 - t)^2 + 3d_2 t^2(1 - t) + d_3 t^3] \right.$$

$$\left. \times \frac{d}{dt} [c_0(1 - t)^3 + 3c_1 t(1 - t)^2 + 3c_2 t^2(1 - t) + c_3 t^3] \right\} dt$$

$$(4.36)$$

This integral can be evaluated to provide an equation for the area directly in terms of the coefficients.

The length of the cubic segments can be computed by using the expression (4.10) for the length of a curve.

### Surfaces

In the design of most engineering products, the *surfaces* of solid objects have to be specified. Fortunately, the idea of the Bézier parametric cubic curve segment can be extended to what is called a *bicubic patch* (see Fig. 4.13), defined by a vector equation in terms of two parameters, $u$ and $v$. The patch has 16 control points, as opposed to only 4 points for the line segment, by which the surface is described.

Since two coordinates (or parameters) are needed to define a point on a surface, the parametric equation for a point $r(x, y, z)$ on the surface must be described in terms of the vector position of the 'control points' and the two parameters $u$ and $v$, which range from $(0,0)$ to $(1,1)$ over the patch:

$$
\begin{aligned}
r = {} & P_{00}(1 - u)^3(1 - v)^3 + 3P_{01}(1 - u)^3v(1 - v)^2 \\
& + 3P_{02}(1 - u)^3v^2(1 - v) + P_{03}(1 - u)^3v^3 \\
& + 3P_{10}u(1 - u)^2(1 - v)^3 + 9P_{11}u(1 - u)^2v(1 - v)^2 + \cdots \\
& + P_{33}u^3v^3
\end{aligned}
\tag{4.37}
$$

An arbitrary surface can be described by placing patches edge to edge to form a *patchwork surface*, as in Fig. 4.14. Continuity of the surface and continuity of the *slope* (in two directions) at the boundaries between the patches is ensured by matching or aligning

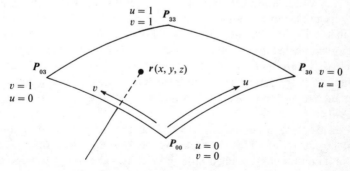

*Figure 4.13.* An element of a surface.

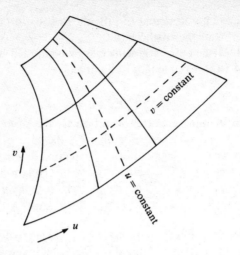

*Figure 4.14.* A surface divided into a patchwork of curved
elements.

the control points. The technique used is much the same as for
curves, but continuity across the boundaries is much more difficult
to achieve. In fact, exact continuity is usually impossible to achieve
with the 'simple' bicubic patch scheme described here. Commercial
surface patch design systems use more complicated patch equations
to achieve continuity, but the general principles are the same.

### From CAD to CAM

Once the surface of an object is represented as a patchwork, and the
coordinates of each point on the surface are known in terms of the
parameters $u$ and $v$, the engineering designer can use the information
in many ways. Computations about the surface (area, volume, line
lengths, etc.), and also about engineering properties of the object, can
be made. Design changes are made to meet as many requirements
as possible (whether the object will do its job properly, its cost, style
and ease of production, and so on). The next step is to transfer the
definition of the surface stored in the computer to the manufacturing
system. We are now in the realm of computer-aided manufacturing
(CAM).

For example, a curved surface such as the blade of a jet engine
may be cut out of a block of metal by the rotating cutters (of radius
$r_c$) of a computer-controlled milling machine. The computer has to
bring the cutter down to within a distance $r_c$, *perpendicular* to the

surface. It has to compute this position and give instructions to the cutter on how to move over the surface, and how rapidly it can produce the final surface. The parameters $u$ and $v$ are very helpful in defining all these instructions.

### Measuring floods and transmitting photographs – approximations of curves with steps and spikes

We have seen how to calculate the area enclosed by curves and the lengths of curves, and how to use these calculations to help *choose* functions that make the area large or small compared with the length. Now we look at functions that might arise in design from using simpler 'piecewise' functions that smoothly join onto each other.

We want to find ways of approximating functions by other functions that are even simpler than the straight-line segments we used to estimate integrals, or the cubic curves we have just considered. In fact, we want to approximate smooth continuous functions by discontinuous functions. It sounds as if our mathematics is going from bad to worse!

#### *Floods and steps*

Suppose the weather is doing its worst and we have several hours of rain. The water drains off fields, into streams and then into rivers. You may have noticed gauges near weirs or rivers; they record levels from which the *flow*, $q(t)$ (in $m^3 s^{-1}$), of water into the river can be computed. The graph of $q(t)$ shown in Fig. 4.15 shows how the flow rate rises and falls over several hours after one hour of rain.

Because each pattern of rainfall is different and the readings may not be very accurate, it is usual for hydraulic engineers to plot the volume of water, $V(t_n)$, as a bar chart. This is because the important quantity is not the *rate* of water flowing over a weir at an instant of time, but the *amount* of water, $V_n$ or $V(t_n)$, that flows between the $(n-1)$th and the $n$th hour. Therefore $V(t_n)$ is a *discontinuous function*. It is defined in terms of $q(t)$ by

$$V(t_n) = \int_{n-1}^{n} q(t)\,dt \quad \text{for } n \geqslant 1 \tag{4.38}$$

Usually the step graph similar to $V(t_n)$ is the only one available. (It is called a *hydrograph*.) Suppose you wanted to estimate *between* the steps. For example, you might need to know how much water had flowed every half-hour, call it $V(t_{n/2})$. The best way is first to

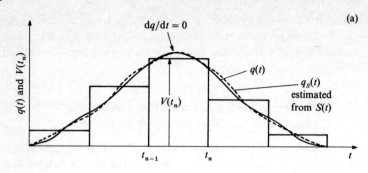

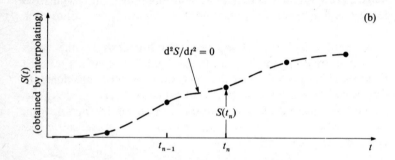

*Figure 4.15.* The flow of water into a river after heavy
rainfall. (a) The true curve $q(t)$ is approximated by a series
of steps, $V(t_n)$. (It is difficult to estimate $q(t)$ from $V(t_n)$.)
(b) Estimating the integral of $q(t)$ from the steps $V(t_n)$ by
interpolating – or guessing – a dashed line between the
points. By differentiating $S(t)$, we estimate $q_S(t)$ – the
dashed line in (a).

estimate $q(t)$ from $V(t_n)$. It would be a bad mistake just to sketch a
line through the discontinuous hydrograph, as you don't know
where $q(t)$ cuts the curves. But you do know the exact value of the
integral of $q$ at the hourly intervals $t_n$ because, from equation (4.38),
this is the *sum* of the hourly values of $V_n$. We call this sum $S(t_n)$:

$$\int_0^{t_n} q(t)\, dt \ = \ \sum_1^n V_n \ = \ S(t_n) \qquad (4.39)$$

We assume there is a smooth function $S(t)$ which is equal to $S(t_n)$ at
the hourly intervals. Then we can interpolate $S(t)$ between points $t_n$
by drawing a curve, and by differentiating $S(t)$ we can estimate $q(t)$

for all $t$. We call this estimate

$$q_S(t) \approx \frac{\mathrm{d}S}{\mathrm{d}t} \qquad (4.40)$$

What we have done, essentially, is calculate a smoothed form of $q(t)$, because any small 'wrinkles' in the graph of $q(t)$ do not affect $S(t)$.

By using differential calculus (which we met in Chapter 1), we can show that when $q_S(t)$ is a maximum, $\mathrm{d}q_S/\mathrm{d}t = 0$ and therefore $\mathrm{d}^2 S/\mathrm{d}t^2 = 0$.

Once the dashed line through the known points in Fig. 4.15(b) is constructed, we could calculate the half-hour values of $V$, which are defined as

$$V(t_{n/2}) = \int_{(n-1)/2}^{n/2} q_S(t)\,\mathrm{d}t \qquad (4.41)$$

This result also shows why integrals of functions are often stored or recorded – because the small errors are then averaged out. For example, if the measurements of the water level are made erratically, any estimate for the missing measurement of $q(t)$ would have less effect on the *integral* of $q(t)$.

### Transmitting pictures and spikes

As a famous BBC cameraman, Stan Spiel, once said, you *usually* don't need to look through a lens to know what the camera sees – you just half-close your eyes and squint! It is rather the same with approximations to mathematical curves.

Look at the photograph in Fig. 4.3(a). Then look at the pictures next to it, Figs. 4.3(b) and (c) – these are the forms in which pictures are transmitted by cable or television. The smooth picture has become a collection of dots. If we let the function $f(x)$ represent the 'greyness' of a photograph, where $f(x) = 0$ for white and $f(x) = 1$ for black, then on a line across the photograph $f(x)$ appears to be a continuous function (see Fig. 4.4(a)). But in the transmitted picture $f(x)$ takes a sequence of values $f_n$, where either $f_n = 0$ or $f_n = 1$, for a sequence of $N$ points over the line: $x = x_n$ for $1 \leqslant n \leqslant N$.

By placing a grid in front of the film, and by suitable developing, a 'half-tone' picture of black and white dots is produced. In typical newspaper pictures there are about 26 dots per centimetre in each direction ($676$ per $cm^2$). In magazines and books, printed on better paper, there may be 40, 60, or even as many as 80 dots per centimetre.

It is straightforward to transmit such a picture by electronic signals, since each point on the picture is denoted by 1 or 0.

With the integrals we have developed, we can now explore mathematically in what *sense* it is possible to approximate $f(x)$ by the sequence $f_n$. Just as, on p. 84, we had to introduce a small length $l_m$ to define what we meant by ingredients being mixed, so now we have to introduce a length $l$ to define a length over which the representation by the dots is satisfactory. By 'satisfactory' we mean that, by 'squinting' (or averaging) over a length $l$, we cannot tell the difference between the two pictures. Since averaging means taking the integral of the function, the total amount of 'blackness' must be the same as the 'area' of the spikes (Fig. 4.4(b)), as $x$ moves a small distance $l$ from $x_1$ to $x_1 + l$: this is equivalent to $m$ dots or spikes (where $m \Delta x = l$). We can equate the integral to the sum:

$$\frac{1}{l} \int_{x_1}^{x_1 + l} f(x)\, dx \; = \; \frac{1}{m} \sum_{n}^{n+m} f_n \Delta x \qquad (4.42)$$

In other words, as $x$ increases to $x + l$, there will be $m$ spikes and gaps (or dots and blanks), but the average amount of 'blackness' must be the same as for the continuous function. Since the effect of any one spike or dot must be *small*, on average the number $m$ of spikes and gaps must be *large* within the distance $l$.

There are many other examples where continuous or 'analogue' signals are converted into this binary (or 0, 1) form. In fact, this procedure leads to more accurate transmission of music and speech.

## Summary

We have shown how the mathematics of curves, lines and areas can be used to choose the best shapes and designs for everyday problems. Our first example was from ancient times. Nowadays, though, the best shapes and designs are usually too complicated to work out with pencil and paper, and so this kind of mathematics is used in conjunction with computers.

## Further reading

R. Courant and H. Robbins, *What is Mathematics?* (Oxford University Press, 1941).

J. M. Ottino, 'The mixing of fluids', in *Scientific American*, Jan. 1989.

# 5

## NETWORKS: UNDERGROUND MAPS, DECISIONS AND AERODYNAMICS

### JULIAN HUNT

### Four examples

#### *Example 1: Undergrounds and networks*

You are explaining the London Underground to a friend who wants to travel from Piccadilly to the Tower of London and then to King's Cross. You could draw a map to represent the lines as they would appear on a street map (see Fig. 5.1(a)). But probably you would draw a map as a set of mostly straight lines which intersect at various places (Fig. 5.1(b)) and include on it extra lines just to show where your friend should not go! Of course, this is the kind of map that is displayed on all Underground stations. Why? Because you need a map so you can make decisions about travelling, not because you want to relate the Underground network to the street network.

There are many different kinds of network that can be represented in an idealized way by a mathematical network of lines which intersect at particular points. Leonhard Euler was the first great mathematician to realize that this kind of map greatly helps in making decisions. On his walks in the old city of Königsberg, he wondered whether he could cross all its bridges without crossing the same bridge twice (see Fig. 5.2(a)).

Nowadays this kind of map is always used for analysing networks of, for example, telephones, roads, railways and electricity grids. In mathematical language, these networks are called *graphs* – but are quite different to the graphs of functions introduced in Chapter 1.

It is obvious that these network diagrams *look* very similar for different applications. In fact there are also mathematical and logical similarities in how they are drawn up and in how they can be used, so it is possible to take the ideas behind one system, and the methods

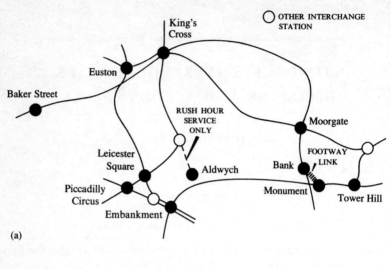

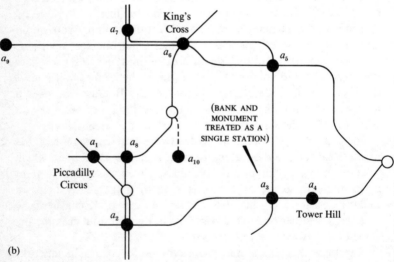

*Figure 5.1.* The London Underground: (a) as a 'real' map
would show it, and (b) as it is usually represented, as a
network, or graph (whose vertices, $a_1$ to $a_{10}$, appear again
in Fig. 5.6). This is just the part of the map required
for a journey from Piccadilly to Tower Hill and then to
King's Cross.

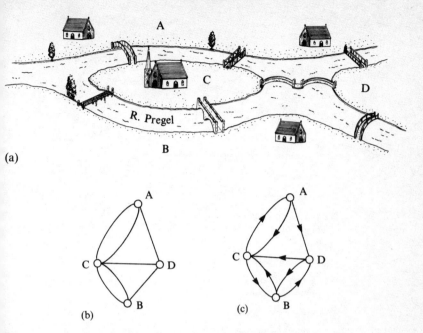

*Figure 5.2.* Leonhard Euler's problem: could he take a walk in the city of Königsberg (a), crossing each bridge once and once only? The graph of the problem (b), in which the vertices A, B, C and D represent the different land areas, shows that such a route is impossible. But with an extra bridge added, as in (c), Euler could have had his walk – one possible route now is ADCACBDBC.

of analysing and operating it, and apply them to other systems. Even quite simple mathematical thinking and calculations can be very useful in both these respects.

### Example 2: Planning a complex operation

Suppose you had to plan the building of a ship. You might start by writing a list of all the different stages: first the design, then the building of the keel, the hull, the engines and the superstructure, and finally things like navigation and communication equipment (see Fig. 5.3(a)). But you would not simply do one operation after another, because it would take far too long. In your plan, the work on the design would start at the same time as steel was being ordered. Then, while steel sheets were being laid for the keel, work would be

| Operation | Time taken (months) | Completion point in graph |
|---|---|---|
| Initial design | 1 | ① |
| Detailed design | 2 | ② |
| Steel ordered | $1\frac{1}{2}$ | ② |
| Keel | 1 | ③ |
| Hull | 2 | ④ |
| Engines ordered and built | 2 | ④ |
| Superstructure built | 2 | ⑤ |
| Engines installed | 2 | ⑤ |
| Navigation and other equipment ordered and built | 2 | ⑥ |
| Superstructure installed | 1 | ⑥ |
| Navigation and other equipment installed | 1 | ⑦ |
| Trials | 1 | ⑧ |
| Internal fittings installed (heating, lighting, etc.) | $1\frac{1}{2}$ | ⑧ |

(a)

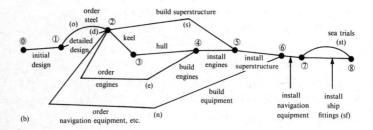

(b)

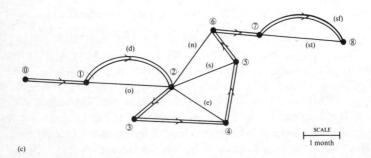

(c)

under way on the hull and superstructure. If several operations taking various lengths of time are happening concurrently, then only when the longest of these steps is completed can the next stage begin. Clearly, if the design is completed, the keel cannot begin to be laid without the steel being ready!

These kinds of planning or operational problem can be understood and solved by drawing networks, or *decision trees*. Thus, in Fig. 5.3(b) we have replaced the list in (a) with a network diagram (graph) to show which operations are carried out at the same time and which in sequence. The completion of each operation is marked by a point, or *vertex*, on the diagram, and is labelled with a number. For example, the moment when the engines have to be finished and the hull completed is given the number 4; then the engines have to be installed by 5, before the superstructure is built on the hull. In making the plan, we would estimate how long each operation is going to take. We can represent the length of an operation by the length of the line between the two vertices, from 4 to 5, say. Can you see from this graph how long the ship takes to build – that is, can you identify the longest, or 'critical' path (see Fig. 5.3(c))?

With some simple mathematics, as we shall show, it is possible to use this kind of graph in several interesting ways.

### Example 3: Aerodynamics

We have seen that the key to understanding Underground maps and complicated sets of decisions is to concentrate on the intersection points, or vertices, of the network. The same approach is now being used to analyse the air flow over many kinds of aircraft wing, such as the swept-back wings of supersonic transport or fighter planes. It is also used for more everyday kinds of flow, such as the flow around

---

*Figure 5.3.* Planning to build a ship. (a) Listing each operation. The ringed numbers correspond to the vertices in parts (b) and (c), points at which operations reach completion. (b) The various operations represented by lines, which meet at points where operations come together. (c) The same diagram, but drawn to scale: the time of an operation is represented by the length of its line. The longest line, from ⓪ to ⑧, is represented as a double line and is the *critical path*: it determines the time of the whole operation of building the ship.

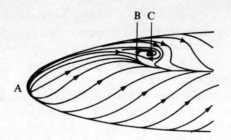

*Figure 5.4.* Pattern of flow lines over the nose of an aircraft
or a rocket.

buildings or along pipes. Figure 5.4 shows the pattern of the flow
lines over part of the surface of the nose of an aircraft or a rocket.
There are places where lines spiral in to a point; elsewhere lines
intersect at 'crossroads'.

These kinds of pattern are usually produced by vortices in the
flow, which are quite similar to those drawn by Leonardo da Vinci
and shown in Fig. 1.3.

Of course the main difference between this pattern of lines and the
lines in the previous networks is that there are infinitely many flow
lines here, but there are only a finite number of lines in the networks.
However, there are only a finite number of points where the lines
cross or converge. These are called *singular points* and, as we shall
see, they can be analysed by using some of the same mathematics we
use for vertices in networks.

### Example 4: Predators and prey

Now change speed, from an aircraft at $300\,\mathrm{m\,s^{-1}}$ to an arctic hare
being chased by a lynx at a few metres per second over the Canadian
tundra. The predator has to catch the prey on enough occasions to
have enough food to survive. On the other hand, the prey has to
escape often enough to be able to breed fast enough for its population to survive. Does this lead to a constant level of population of
hares ($h$) and lynxes ($l$), or to one or the other dying out, or perhaps
to a regular fluctuation in their numbers? Figure 5.5(a) shows that,
over several years, the populations fluctuate: when $h$ increases $l$
decreases, and vice versa. These two graphs can be plotted on a
single graph of $h$ against $l$. The result is a closed curve (Fig. 5.5(b)).

One of the reasons for studying animal populations with mathematical models is to be able to predict what might happen if there
were a dramatic rise or fall in the numbers, caused by disease,

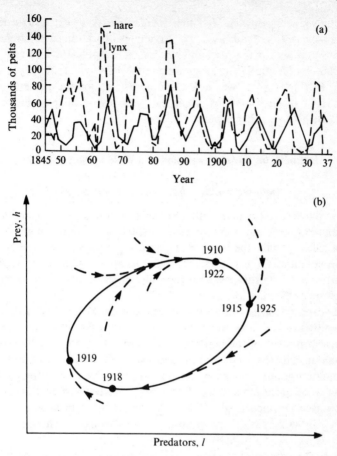

*Figure 5.5.* (a) Variation in the populations of hares and lynxes in the Hudson Bay area of Canada, as represented by the numbers of pelts received by the Hudson Bay Company, 1845–1937. (b) The same data plotted in the 'phase plane'. The ellipse shows how *h* and *l* vary with time; the dashed lines show how the variation would alter if either *h* or *l* were to depart from the equilibrium cycle.

perhaps, or a new form of pollution. The arrows in Fig. 5.5(b) show how the populations of *l* and *h* are likely to change as a result of such 'perturbations'.

These last two examples show how it is possible to analyse continuously changing patterns in *space*, such as flow lines, and also by similar analyses the patterns associated with continuous and complex changes in *time*. In other words, just as we can extend the ideas behind simple railway maps to complex operations with critical-path diagrams for the decisions, we can also apply the analysis of continuous functions and singular points to systems that vary with *time*. In this chapter we look at some aspects of the patterns of behaviour of such systems. Chapter 7, on biological populations, deals with some other aspects.

## Journeys on the Underground and building ships faster

### *Defining graphs, diagrams and networks*

What the layperson might call a network or a pattern of connected lines intersecting at a number of points, like the important stations on a railway map, the mathematician calls a *graph*. We shall need some definitions and notation in order to apply the logic of mathematics to solve problems about graphs. First we give each graph a label: $G_1$ or $G_2$, say, or, in general, $G$.

A graph $G$ consists of a set of special points $V$ called *vertices* (or sometimes *nodes*), say $a_1, a_2, \ldots, a_V$. Figure 5.6 shows a graph representing part of the Underground map in Fig. 5.1. The lines (or curves) joining the vertices are called *edges*. There is an edge joining the second point $a_2$ to the eighth point $a_8$, and we label it $(a_2a_8)$. In order to be quite general, we allow some vertices to be unconnected (such as $a_{10}$, which might represent a station not in service), and some edges to cross others without passing through an intersection, such as $(a_6a_9)$ and $(a_7a_8)$.

In some graphs each edge is defined with a particular sense of direction, like a one-way street, so that it would be possible, perhaps,

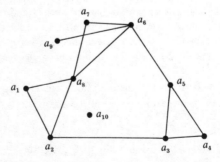

*Figure 5.6.* An even simpler representation of the London Underground maps in Fig. 5.1 – as a graph.

to travel from $a_1$ to $a_2$ but not from $a_2$ to $a_1$. These are *digraphs*. In other graphs, there is no particular direction.

In some graphs there can be more than one line joining a pair of vertices, as in Euler's bridge problem in Fig. 5.2, or the lines between Moorgate and King's Cross in Fig. 5.1.

### *Counting and sorting the edges and vertices*

We shall see that some of the most useful results about the mathematics of networks come from counting the edges and vertices and considering how many there can be. Once a network becomes large and complicated, it is important to know what the general rules must be.

First of all, given the number $V$ of vertices, what is the *maximum* number $E$ of edges there can be? (For this problem we assume that there is only *one* line between a pair of vertices, as in Fig. 5.6. This is a simple example of the kind of question that any communications engineer has to be able to answer. The answer can be calculated by 'induction'. Let the number for $V - 1$ vertices be $E(V - 1)$. So for $V$ vertices, $E(V)$ must be $E(V - 1) + V - 1$, because the new vertex must be connected to all the previous $V - 1$ vertices. Then we calculate $E(V - 1)$ in terms of $E(V - 2)$, and so on, down to $E(2)$. $E(1)$ is zero. This leads to

$$E(3) = E(2) + 2$$
$$E(4) = E(2) + 2 + 3 \qquad (5.1)$$
$$E(5) = E(2) + 2 + 3 + 4$$

So

$$E(V) = \sum_{r=2}^{V-1} r + E(2) \qquad (5.2)$$

Since for a two-vertex graph $E(2) = 1$, and since $2 + 3 + 4 + \cdots + V - 1 = (V - 2)(V - 1 + 2)/2$, it follows that the maximum number of edges is

$$E(V) = \sum_{r=2}^{V-1} r + E(2) = \frac{(V - 2)(V + 1)}{2} + 1$$
$$= V(V - 1)/2 \qquad (5.3)$$

We see that the maximum number of edges increases rapidly, in fact as the square of the number of vertices. For $V = 10$, in Fig. 5.6, the actual number of edges is 12, whereas the *maximum* possible number of edges or lines is $(8 \times 11)/2 + 1 = 45$.

The number of edges $E$ can also be related to the 'importance' or *degree* of the vertices. Just as you might define the importance of a railway station, or an airport, by the number of lines passing through it, mathematicians define the degree of a vertex as the number of edges that end at that vertex. In Fig. 5.6, at the vertex $a_2$ there are three edges, so its degree is 3, which we denote by $d_2 = 3$. At the vertex $a_9$ there is one edge, so $d_9 = 1$. There is no edge at the isolated vertex $a_{10}$, so $d_{10} = 0$.

By counting all the edges and noting that each edge joins two vertices, it follows that the sum of all the degrees of the vertices is equal to *twice* the total number of edges $E$:

$$\sum_{i=1}^{V} d_i = 2E \qquad (5.4)$$

For the graph in Fig. 5.6 this sum is equal to $2 + 3 + 3 + 2 + 3 + 4 + 2 + 4 + 1 + 0 = 24$, and there are 12 edges, so $2E = 24$, which verifies equation (5.4). This sequence of numbers shows that there are six vertices whose degree is an even number, and four whose degree is odd. The sum of $d_i$ of all the even vertices (14 in this case) is an even number. Therefore, since $2E$ is an even number, the sum of the degrees of the odd vertices must also be an even number (in this case $d_1 + d_3 + d_4 + d_5 = 10$). But the degree of each odd vertex is *odd*. If a sum of odd numbers is an even number, then there must be an even number of them (in this case there are four vertices with odd degrees – or *odd vertices*).

Therefore we have shown that there must always be an even number of vertices at which there are an odd number of edges.

*Finding the best routes for delivery vehicles*
*Avoiding the same route a second time*

It is the degrees of the vertices of a graph that provide the clue to understanding Euler's problem of crossing the Königsberg bridges, and to other problems such as organizing the best routes for sales representatives or delivery vehicles.

With Fig. 5.2, the question was to pass along each bridge (or edge) once, without recrossing any bridge. We can recast the problem in terms of the graph in Fig. 5.2(b): we must start from any one of the vertices A, B, C and D, and go along each edge once and once only.

(Note that we have put the whole left bank at B, and the right bank at A, because we are interested only in the bridges, not in routes along the banks or across the islands.) The path must end at a vertex, so there must be at least two intermediate vertices on the route where the path does not end. If there is to be no recrossing of a bridge, there must be an even number of bridges connected to one of these intermediate vertices. *But the degree of every vertex is odd.* Therefore there is no route which starts at A, B, C or D, and ends at A, B, C or D, and passes over each bridge once. Try for yourself!

A similar problem would arise if you had to organize a delivery round along routes between a number of points, $a_1$ to $a_n$, say. You would want to know whether it were possible to do this without the delivery vehicle having to travel along the same roads twice. Rather than answer this by trial and error, you could draw a graph of the routes and then note which of its vertices were even and which were odd. As we saw with Euler's bridge problem, it follows that it is possible to avoid passing over the same point twice only if the degrees of the intermediate vertices of the network $a_2, \ldots, a_{n-1}$ are *even*. At the beginning and end points, the degrees of the vertices can be odd!

For example, let the graph in Fig. 5.2(c) represents the roads linking the four delivery points A, B, C and D. This is the same as Fig. 5.2(b), but with one extra edge, between D and B. Can you see at which points it is possible to start and end a non-repeating route? Now the degrees of B and D are even (so it is possible to start at A and end at C, or vice versa).

## How many different routes?

In many network problems, from telecommunications to transportation, it is necessary to know the *number* of possible routes from one vertex of the network to another when the route passes through intermediate vertices. Here we shall use some mathematics introduced in other chapters in this book, such as matrices and suffix notation from Chapter 2.

First, we label every vertex (as in Fig. 5.6). We use the notation $a_{ij}$ to denote the direct path from the $i$th vertex to the $j$th vertex, and $a_{ji}$ to denote the direct path from the $j$th vertex to the $i$th vertex. (In this analysis there can be only a single path in one direction between two vertices.) If such a direct path exists, then $a_{ij} = 1$, and if it does not, then $a_{ij} = 0$.

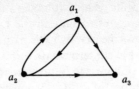

*Figure 5.7.* A simple directed graph.

In the simple example shown in Fig. 5.7, there are paths both ways between $a_1$ and $a_2$, but in only one direction between $a_1$ and $a_3$ and between $a_2$ and $a_3$. There are no paths *from* $a_3$. Therefore we can arrange the $a_{ij}$ as a matrix $A$, where $a_{ij}$ is the $(i,j)$th element (i.e. the element in the $i$th row and the $j$th column):

$$A = \begin{pmatrix} 0 & 1 & 1 \\ 1 & 0 & 1 \\ 0 & 0 & 0 \end{pmatrix} \tag{5.5}$$

Since there is no path from $a_1$ to $a_1$, the matrix element $a_{11}$ is zero; similarly, $a_{22} = a_{33} = 0$.

We can use this notation to calculate the number of paths consisting of one or two edges, $P_{ij}^{(2)}$, from the vertex $a_i$ to the vertex $a_j$, either directly or passing through only *one* intermediate vertex. For example, how many ways could you travel from Piccadilly to King's Cross? Or, in the simpler case of Fig. 5.7, how many ways are there of travelling from $a_1$ to $a_3$ either directly or via one intermediate point?

The solution is found by considering the paths $a_{ik}$ from $a_i$ to *all* the other vertices, which we label as $a_k$ (for any $k$, if $k = i$, $a_{ik} = 0$). These are the intermediate vertices. We then consider the paths $a_{kj}$ from each vertex $a_k$ to the vertex $a_j$. If both paths exist, from $a_i$ to $a_k$ and from $a_k$ to $a_j$, then $a_{ik} = 1$ and $a_{kj} = 1$, and also $a_{ik}a_{kj} = 1$. If one or other does not exist, then $a_{ik}a_{kj} = 0$. Therefore, by summing $a_{ik}a_{kj}$ over all values of $k$, we have found the total number of paths from $a_i$ to $a_j$ passing through one intermediate point.

But $a_{ik}a_{kj}$ is equal to the $(i, j)$th point in the matrix obtained by multiplying the matrix $A$ by itself, i.e. by squaring $A$. So the number of paths $P_{ij}^{(2)}$ from $a_i$ to $a_j$, either direct or via one intermediate point, is given by

$$P_{ij}^{(2)} = a_{ij} + \sum_{k=1}^{3} a_{ik}a_{kj} \tag{5.6}$$

So, in our simple example in Fig. 5.7,

$$P_{13}^{(2)} = a_{13} + (a_{11}a_{13} + a_{12}a_{23} + a_{13}a_{33})$$
$$= 1 + (0 + 1 + 0) = 2 \tag{5.7}$$

(remember that $a_{11} = a_{33} = 0$).

We could have squared the matrix $A$ to obtain $P_{ij}^{(2)}$ from the $(i, j)$th element of the matrix $P^{(2)}$, where

$$(P^{(2)}) = A + A^2$$

$$= \begin{pmatrix} 0 & 1 & 1 \\ 1 & 0 & 1 \\ 0 & 0 & 0 \end{pmatrix} + \begin{pmatrix} 0 & 1 & 1 \\ 1 & 0 & 1 \\ 0 & 0 & 0 \end{pmatrix} \begin{pmatrix} 0 & 1 & 1 \\ 1 & 0 & 1 \\ 0 & 0 & 0 \end{pmatrix}$$

$$= \begin{pmatrix} 0 & 1 & 1 \\ 1 & 0 & 1 \\ 0 & 0 & 0 \end{pmatrix} + \begin{pmatrix} 1 & 0 & 1 \\ 0 & 1 & 1 \\ 0 & 0 & 0 \end{pmatrix}$$

$$= \begin{pmatrix} 1 & 1 & 2 \\ 1 & 1 & 2 \\ 0 & 0 & 0 \end{pmatrix} \tag{5.8}$$

### Networks on surfaces

Mathematical analysis of networks, or graphs, tells us about connections between vertices, such as different routes between stations on the Underground, or links in a telephone system. But when graphs are related to real lines on real surfaces, such as a supply network, or metal crystals, or flow lines on an aircraft wing (Fig. 5.4), we find that lines do not cross each other, except at special points where many lines come together or diverge in different directions. The mathematical rules are now different, and the mathematical analysis can yield important results about the relations between lines and the spaces between them.

Consider a network of edges connecting vertices drawn on a plane, such as those we looked at in the previous section. These edges divide the plane into *regions*. Individual regions correspond to, say, individual faces of a crystal or countries on a map.

It is possible to travel anywhere within a region without crossing any edges of the graph. No lines cross, except at the vertices. There

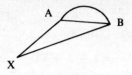

*Figure 5.8*. Extending a graph. For each new vertex (X) there
are two new edges (AX and BX) and one new region (AXB).

is also a region outside the graph, consisting of everywhere in the
plane not enclosed by the edges of the graph. (We exclude networks
with unconnected vertices; it must be possible to move around the
network to all points.)

We have seen in equation (5.4) that there is a relation between the
number of edges $E$ of a network on a plane and the sum of the
degrees of all the vertices. This sum equals the total number of lines
meeting at the vertices. So this must tell us something about how the
number $R$ of regions is related to the numbers of vertices and edges.

What happens if we add two edges to a graph, from A and B to
X (see Fig. 5.8), by drawing the two edges from existing vertices?
This creates *one* new vertex (X), where the edges meet, and *one* new
region (AXB). We can construct any kind of graph on a plane by
adding more edges and vertices. So the number of new regions
created is always equal to the number of new edges less the number
of new vertices.

This explains the remarkable result that, for any graph on a plane,
there is a unique relation between the total number of vertices $V$, the
total number of edges $E$, and the total number of regions $R$:

$$V - E + R = G \qquad (5.9)$$

The constant $G$ is 2. (Just think of the simplest two-vertex, two-edge,
two-region network, and you will see why.) In this case equation
(5.9) is known as *Euler's theorem*. One very simple but important use
of this formula is for checking that a diagram consisting of a pattern
of shapes, such as a section of metal crystals, has been correctly
drawn. But it also provides the basis for various calculations about
networks and maps.

Suppose you wanted to supply each of three chemical reactors A,
B and C with three reactants D, E and F (or supply each of three
houses with electricity, gas and water), as shown in Fig. 5.9(a). But
you want to know whether it is possible to do this *without* the supply
lines crossing each other. Drawing the graph (Fig. 5.9(b)) shows that
it is not possible.

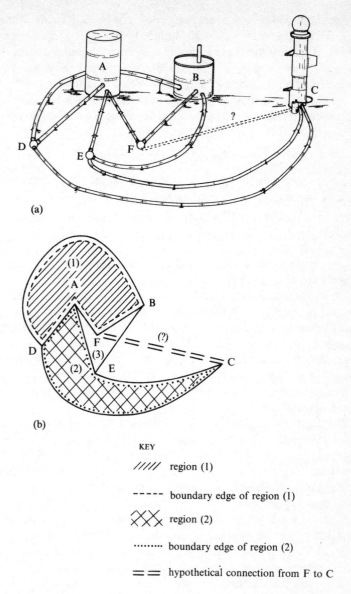

(a)

(b)

KEY

///// region (1)

----- boundary edge of region (1)

XXX region (2)

········ boundary edge of region (2)

== hypothetical connection from F to C

*Figure 5.9.* (a) A system of pipes in a chemical plant
connecting three reactor vessels A, B and C with three
sources of reactant D, E and F. Is it possible to complete the
network by a pipe joining C to F that does not pass under
or over other pipes? (b) Representing the problem by a
graph.

For more complex networks, we need a method of analysis that does not require us to draw many graphs by trial and error. Let each region of the network have a certain number of boundary edges. In Fig. 5.9(b) the region within ADBFA has four edges; AD is a boundary edge of ADBFA and of ADCEA. The total number of edges is $E_R$ (these may be counted twice). Since in this particular network each region has four or more sides (there would be no point in joining A to B or E to F), it follows that

$$E_R \geqslant 4R \qquad\qquad (5.10)$$

But since each actual edge may or may not count as the boundary edge of two neighbouring regions, the total number of actual edges is greater than $E_R/2$, and therefore

$$E \geqslant E_R/2 \qquad\qquad (5.11)$$

In our imaginary case, where A, B and C are connected to D, E and F, we have $E = 9$, so $E_R \leqslant 18$.

Therefore, combining the two conditions (5.10) and (5.11), we find that the restriction on $R$, the number of regions, is $R \leqslant 18/4$. But by Euler's theorem, equation (5.9), $R = 2 + 9 - 6 = 5$. Therefore since it is impossible for $R$ to be equal to 5 and less than $4\frac{1}{2}$, our network cannot be non-overlapping. So one reactor would remain without one of the reactants, or one house would have to go without one of the services – or one of the service pipes would have to go under another one!

The main reason why mathematicians have been interested in Euler's theorem is that the constant $G$ changes when the vertices and edges of the network are distributed over different *shapes* (Fig. 5.10). But what began as mathematical fun has led on to important mathematical theories in topology. It also leads to interesting and important applications.

If you draw a graph, or 'map', on the surface of a *sphere*, the constant $G$ remains the same. (In Fig. 5.10, $V = 2$, $E = 3$ and $R = 3$.) This applies to any shape that can be deformed into a sphere without making holes in it. Therefore Euler's theorem applies to the vertices and edges of any solid shape, such as a tetrahedron ($V = 4$, $E = 6$, $R = 4$), or a dodecahedron (where each face (region) has five sides, and $V = 20$, $E = 30$, $R = 12$).

But what happens if the network is drawn on a torus, in such a way that it could *not* be drawn on a spherical or a plane surface? Figure 5.11(a) shows a network of lines with two edges and one vertex, at A, but just *one* region. In other words, it is possible to

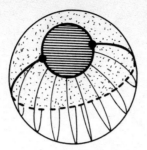

*Figure 5.10.* A graph on a sphere. It has two vertices, three edges and three regions.

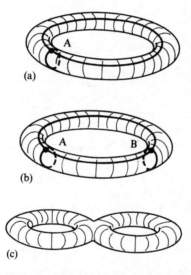

*Figure 5.11.* (a) A graph on a torus, with one vertex (A), two edges and one region. (b) With one extra vertex (at B), there are now four edges and two regions. (c) A double torus.

move everywhere on the surface of the torus without crossing an edge. Figure 5.11(b) shows how a second, unconnected region can be added by introducing another vertex at B, and another ring around the torus. Then $E = 4$, $V = 2$ and $R = 2$. So in both these cases, and for any other network over the torus, $V - E + R = G = 0$.

This constant $G$ is called the *Euler characteristic* of the surface. It is 2 for any shape that can be deformed into a sphere, 0 for any shape that can be deformed into a torus, $-2$ for a double torus (see

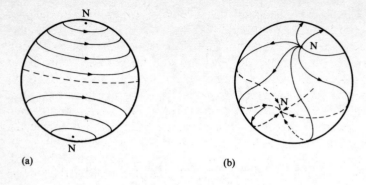

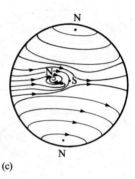

*Figure 5.12.* Flow lines on a sphere. (a) Flow around two
nodes; (b) flow from one node to another. (c) The wind
pattern of a planet.

Fig. 5.11(c)), and so on. Different shapes with the same Euler
characteristic are called *topologically similar*.

The idea that there are essential similarities between patterns of
lines on topologically similar shapes can be applied in many new
ways. In aerodynamics, for example, we need to consider the pattern
of flow lines *everywhere*. There are many other examples of
continuous patterns of directed lines, or *vectors*, over different
surfaces – from the hair on an animal to magnetic fields and electric
currents.

Imagine brushing a hair-covered sphere, so as to give it the
smoothest possible 'hairstyle'. You will always end up with two
points where the hairs have no direction: either 'crowns' where the
hairs go round a point (Fig. 5.12(a)), or 'tufts' where the hairs
converge to or diverge from a point (Fig. 5.12(b)).

If you were to find more than two points where the hair could not be brushed, you would find that there also had to be *other* points where the hairs had no direction. But these points would not look like crowns or tufts. They would look like 'cross points' or, in mathematical jargon, 'saddle points', where some hairs point inwards and some point outwards. (Try drawing three tufts on a sphere – you can't!) This is illustrated in Fig. 5.12(c), a schematic diagram of a planet's wind pattern (the point marked N near the centre of the disc is the extra tuft, and S is the saddle point).

But what happens on a hairy torus? The hairs can be brushed, either round the ring or round the section (directions marked by the two edges in Fig. 5.11(a)), but in neither case are there any points where the hairs have no direction. These special points are called *singular points*. Defining each one of these and their connections really defines the general pattern of the vectors everywhere, just like defining a graph in terms of its vertices and their connecting edges.

These singular points fall into one of two categories – node points (N) or saddle points (S). At the node points shown in Fig. 5.13(a), all the vector lines converge, or diverge, or just circulate round the point. At the saddle points in Fig. 5.13(b), most of the vector lines near the point do not pass through the point. But two 'critical' lines enter the point and two leave it. (For most real flows or other situations, there are two straight lines which cross at the saddle point.)

Why are these different kinds of node point all in one category? One answer is that you can find two of them on a hairy sphere without a saddle point! A more mathematical answer is that if a point travels anticlockwise around any node point, the *direction* of the vector defined by an angle $\theta$ to a fixed direction at each point increases by 360° (see Fig. 5.13(a)). (This is quite useful because it means that the nature of the point can be found without looking very carefully at the hairs or vectors at the point itself!) Around a saddle point, $\theta$ *decreases* by 360° (see Fig. 5.13(b)).

As we saw with the hairy sphere, the total number of node points, $N$, can be 2. But if there are three node points ($N = 3$), there must be one saddle point. In other words, there must be a formula, rather like equation (5.9) for the vertices and edges of a network, but relating $N$, the number of node points, to $S$, the number of saddle points. As $S$ increases from 0 to 1, $N$ increases from 2 to 3, so the formula must involve subtracting $S$ from $N$. In fact,

$$N - S = G \qquad (5.12)$$

where, as with Euler's theorem, the constant $G = 2$.

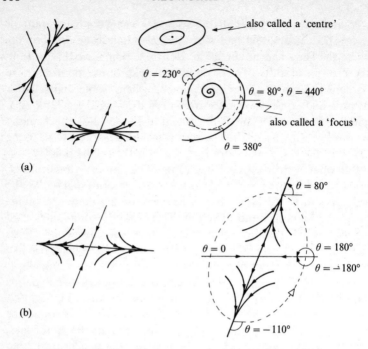

*Figure 5.13*. The two topological varieties of singular point:
(a) node points and (b) saddle points. The change in the
angle $\theta$ of the direction of the vectors around an anticlockwise loop is indicated by a dotted line and arrowhead.

Now think again about the hairy torus. There were no singular
points, so $N = S = 0$. Therefore, as with the network on the torus
(Fig. 5.11), $G = 0$. So for every node point on the surface of a torus
there must be a saddle point. On a double torus $G = -2$, so for
every node point there must be three saddle points, and if there are
no nodes there are two saddle points.

The classification and counting of singular points is now a
standard practice in describing and analysing complicated flows in
aerodynamics, such as the air flow over aircraft wings or around
buildings (with or without passage-ways underneath), or flows at
junctions in pipelines or arteries.

We can consider the aircraft nose in Fig. 5.4 as part of a sphere-
like shape. The air flow impacts the surface of the nose at A, from
which point all the surface streamlines diverge, so A is a node. There
is a 'focus' at C, where the streamlines converge, and a saddle point
at B, where two lines converge and two lines diverge. Thus $N = 2$

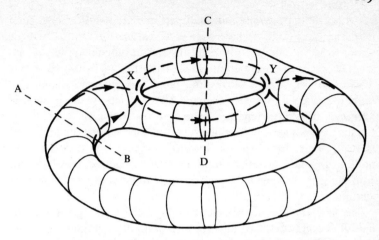

*Figure 5.14.* Streamlines on the surface of a closed pipe,
showing saddle points at X and Y. This arrangement, which
is topologically equivalent to a double torus (Fig. 5.11(c)),
could represent an artery. The section between AB and CD
could represent a single junction in a pipe or artery.

and $S = 1$. Thus the formula (5.12) for the sphere is satisfied —
provided we assume there is another node on the part of the sphere
not shown here. So now $N = 3$ and $S = 1$.

These singular paths are not only important for defining the flow
round them, but they are also important in themselves because there
is *no flow* at these points. At these points there could be a build-up
of solid matter or scaling in pipes in chemical plant or in hot water
pipes, or a build-up of fat in arteries. Figure 5.14 shows two junctions in a closed pipe. Topologically the surface of this pipe is
a double torus, so that

$$N - S = G = -2 \qquad (5.13)$$

Therefore there must be *at least* two singular saddle points, as shown
on the diagram at X and Y, where the flow stagnates. The part of the
diagram between AB and CD shows that, where a pipe divides, there
must be at least one saddle point (at X).

## Behaviour of predators and other time-varying systems

Drawing patterns of lines need not be restricted to *actual* directions
of actual quantities. We saw in Example 4 at the beginning of this
chapter that the variation in the populations of arctic lynxes and

hares can be represented and analysed by patterns of vectors on a diagram. Now that we have seen how patterns of vectors can be understood and analysed by focusing on the singular points where the vectors have no direction, we can use this idea to look at the broad features of how lynxes and hares – or enzyme systems, or other time-varying systems – behave. We can even use this geometric approach to explore 'chaotic' behaviour, as shown for example by the Earth's magnetic field, which reverses itself – north pole flipping to south, and vice versa – at intervals ranging from thousands to millions of years. Another example is the human heart, which can stop suddenly, or start to beat unsteadily, after many years of regular beating.

The first step is to draw a curve on a graph (meaning a 'plot', not one of the networks we have been looking at) representing some aspects of the behaviour of the system at each moment of time. This might be a plot of the number of lynxes ($l$) against the number of hares ($h$), as in Fig. 5.5(b). (Such 'predator–prey' systems are discussed in Chapter 7.)

In other systems there might be just one key variable, such as the flow of blood, $F$, into the heart, or the strength of a magnetic field, $H$. Figure 5.15 shows approximately how $F$ and $H$ might vary with time. These graphs look rather different from those of the lynx–hare system. We also consider how regular oscillations in these systems might arise by using what are called phase-plane diagrams: Fig. 5.16 shows graphs of $dF/dt$ against $F$ and $dH/dt$ against $H$, for different starting values of $F$ and $H$ (not on the regular oscillations). For a large starting value such systems would converge to one or other of the equilibrium oscillations, or to the closed loops, on the diagram. Equally, for small starting values the paths would also tend to converge to the closed loops.

By representing such systems as plots on a phase-plane diagram, many possible kinds of behaviour are easily seen. Some *impossible* kinds of behaviour can be eliminated by applying the topological rules for singular points on the lines of the diagram.

In Fig. 5.16 the vectors (i.e. lines outside the closed loops) are all inwards: $F$ and $dF/dt$ are both directed towards the loops. These vectors are near closed loops similar to those in Fig. 5.13(a), directed inwards towards a node point. If there are also saddle points, the only way topologically for the lines to be directed inwards far from the closed loops is for there to be more node points than saddle points in the plane, i.e.

$$N - S = 1 \tag{5.14}$$

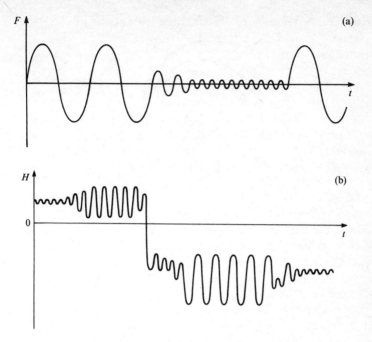

*Figure 5.15.* Two oscillating systems: the time variation of (a) blood flow $F$ in the human heart, and (b) the Earth's magnetic field $H$. The human heart can suddenly start to oscillate faster; the Earth's magnetic field can suddenly 'flip' from one mean state to the other.

(Or you can think of this plane as the top half of a hairy sphere where the vectors are converging to a tuft near the origin!) Therefore with nodes in the two circular loops ($N = 2$) there has to be one saddle point S between them ($S = 1$).

These diagrams have not described or explained how the systems flip from one state to another. This would require the system to move along paths (shown dashed in Fig. 5.16) which cross the other trajectories. Therefore a complete map of the behaviour of the system cannot be represented in two dimensions on a plane. You can read about this in books on chaos and dynamical systems.

## Summary

You could say that mathematics takes ordinary ideas, makes them clear and logical, turns them into symbols, and then works out the

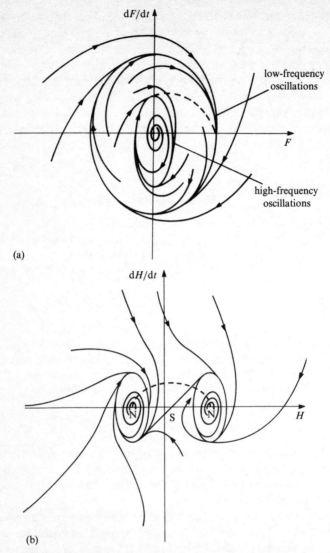

*Figure 5.16.* The variations shown in Fig. 5.15 plotted as phase-plane diagrams. (a) The arrows show how the heart can move from high- to low-frequency oscillation, along time trajectories of changing $F$ and $dF/dt$. Again, the dashed line indicates a flip from one pattern of oscillation to the other.

(b) The Earth's magnetic field tends to one state or the other; the arrows show the time trajectories of changing $H$ and $dH/dt$. The dashed line gives an idea of how the system might 'flip', although this cannot be represented properly on the diagram.

consequences. In this chapter the simple idea was that to use the London Underground you need to know where and how to change trains. This is more important than what happens between the change-over stations. This idea led to the Underground map, which is one example of a network or graph. We have seen that the rules for a network of lines and curves can then be used to help in the planning of business and industry, and to analyse and understand the changes that occur in many complex scientific, technical and medical systems, without ever having to know the details of any of them!

## Further reading

G. Chartrand, *Introducing Graph Theory* (Dover, 1977).
H. G. Flegg, *From Geometry to Topology* (English Universities Press, 1974).
J. Gleick, *Chaos: Making a New Science* (Cardinal, 1988).
I. Stewart, *Does God Play Dice? The Mathematics of Chaos* (Penguin, 1990).

# 6

## SCALING METHODS IN PHYSICS AND BIOLOGY

### JOHN RALLISON

### Scaling the heights

How much would it cost to build a house of twice the size? How much faster will a car travel if its engine capacity is trebled, and how much extra fuel will it consume? If a chemical reactor is halved in size, how much faster does the chemical reaction take place? What is the increase in the destructive power of an atomic bomb of twice the mass? If an adult gazelle is four times the size of its offspring, how much higher can it jump and how much faster can it run?

Questions such as these, about the relationship between some (usually physical) property of interest and size, are matters of *scale*. They are often difficult to answer, in that they may concern the interactions of many complex mechanisms, some of which are only imperfectly understood. Where the questions are important, in economics or engineering, for example, it may be vital to supply answers to them, albeit approximate ones. The power and usefulness of scaling methods is that they can often provide (even if detailed mechanisms are not known) the sort of approximate results desired, for a minimum of analytical or numerical effort.

In simple cases (some of which are discussed below) physical relationships may be known, and scaling methods can then produce answers which are in principle exact. In other cases, a set of observations of a phenomenon (say the maximum swimming speed of different species of fish) over a range of scales may have been made, and the objective is to find the physical mechanism underlying the observations. Scaling methods often enable the immediate rejection of various hypotheses as being inconsistent with the observations – but cannot make it possible to choose between different hypotheses which share the same scaling behaviour. In short, scaling methods can often provide powerful insights into difficult problems, but rarely a full solution to them. This chapter attempts to show, by reference to a number of different physical and biological systems, both the power and the limitations of some of the newer scaling applications, including recent work using *fractals*. We start, however, with a brief review of the basic ideas.

## Geometrical similarity and scaling

The total quantity of leather in a pair of shoes is self-evidently twice as great as in a single shoe. A relationship between two quantities such as these, in which the doubling of one quantity (the number of shoes) doubles the other (amount of leather), is called *linear*, because a graph of one quantity plotted against the other is a straight line, and is the simplest – and probably the commonest – scaling behaviour. (The question of exactly how much leather is required for each shoe is much harder, and cannot be answered by scaling ideas alone.)

Linear scalings lie at the heart of trigonometry. If two plane geometric figures are similar (i.e. have the same angles) then the lengths of the sides of the two figures are linearly related. Thus if one pair of corresponding sides have lengths $l$ and $2l$ respectively, then all other pairs of corresponding sides are in the same $1 : 2$ ratio. One application of this principle is that the length of shadow cast by a post is proportional to the height of the post. Thus (see Fig. 6.1) if a post known to be 6 m tall is found to cast a shadow 4 m long, then a radio mast which casts a 20 m shadow must be 30 m tall. (Sherlock Holmes gives Dr Watson an insultingly full explanation of this idea in *The Musgrave Ritual*.) The correspondence of the angles of the two figures is of course crucial: the length of the shadow changes with the position of the Sun, so both shadow lengths must be measured at the same time of day.

Not all scaling relationships are linear. Perhaps the simplest non-linear example is given by area. The area of a triangle is well known to be $\frac{1}{2}$base × height. The right-hand triangle in Fig. 6.1 has both base and height five times as large as those of the triangle on the left, and thus its area is 25 times greater. This is an example of the general principle that for geometrically similar figures whose lengths scale with $l$, the areas scale with $l \times l = l^2$.

This same scaling idea determines the units in which area is measured. If the unit of length is the metre, that for area will be the square metre (or 'metre squared' in a more natural notation). If the

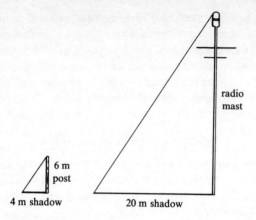

*Figure 6.1.* Similar triangles: a post 6 m high casts a 4 m shadow; then, if a radio mast casts a 20 m shadow (at the same time of day), we can deduce that it is 30 m tall.

unit of length is changed to the centimetre (one hundred times smaller) then that for area (square centimetre) will be ten thousand times smaller. More generally, if the unit of length is written as [L], then that for area is written as $[L]^2$.

Just as lengths scale as L, and areas as $L^2$, so volumes scale as $L^3$. Thus if the radius and height of a cylindrical hot water tank are each doubled, the total volume of water required to fill it will be increased by a factor of eight. (We do *not* need to know the formula for the volume of the cylinder to work this out!) The unit of volume is written as $[L]^3$.

An old conundrum embodying these geometric scaling principles concerns a matchbox manufacturer who decided to produce boxes of matches whose linear dimensions were 50% larger than those of his rivals. After a few months he was inundated with complaints by members of the public. Why? Whereas the number of matches in the box had been increased by a factor of $(\frac{3}{2})^3 = 3\frac{3}{8}$, the area of sandpaper on which the matches were struck had been increased by the smaller factor $(\frac{3}{2})^2 = 2\frac{1}{4}$, and so was wearing out before all the matches could be used.

### Dimensional analysis in physics and biology

Simple geometric scaling can, when combined with physical laws, provide results of great generality and power. We have seen, for example, that the volume of a solid (or fluid) of linear size *l* scales as

$l^3$. If the material has uniform density, then, since mass = density × volume, the mass (and hence the weight) of the material also scales as $l^3$. More generally, if the pattern of density variation between geometrically similar pieces of material is the same, then their masses scale as $l^3$. For instance, if a mature tree is ten times taller than a sapling (and is geometrically similar to it), the weight that must be supported by its trunk is a thousand times larger.

For a body of mass $m$ in motion, we may similarly determine the scaling for its kinetic energy and momentum. If it is travelling at a speed $v$, then its energy $\frac{1}{2}mv^2$ and momentum $mv$ are both proportional to $m$, and hence scale as $L^3$. So if the relationship between physical quantities is known, it is straightforward to determine the corresponding scaling behaviours.

But even when physical laws are not known with precision, scaling ideas can often produce dramatic short-cuts to identifying them. There are two principles underlying this method of *dimensional analysis*:

(1) The statement of a physical relationship does not depend on the particular set of units used to express it.
(2) The units of mass [M], length [L] and time [T] are fundamental in the sense that none can be written in terms of the others.

The first principle can be illustrated by our formula for kinetic energy ($\frac{1}{2}mv^2$). In c.g.s. units mass is measured in grams, and speed in centimetres per second. The unit of energy, the erg, is therefore $1\,\text{erg} = 1\,\text{g cm}^2\text{s}^{-2}$. In SI units mass is measured in kilograms, and speed in metres per second. The unit of energy, the joule (J), is then $1\,\text{J} = 1\,\text{kg m}^2\text{s}^{-2}$ ($= 10^7\,\text{erg}$). In general, then, the unit of energy is [M] [L]$^2$ [T]$^{-2}$; similarly, we write the dimensions of energy as $M\,L^2\,T^{-2}$.

Where a number of quantities whose units are known are supposed to enter into a physical relationship, but the precise relationship is not known, the principles above can often give a short-cut to the scalings in the answer. Consider, for example, a body of uniform density $\rho$ and linear size $l$, travelling at a speed $v$. If we *know* that its energy has dimensions $M\,L^2\,T^{-2}$, and make the intelligent guess that this energy depends *only* on the density $\rho$ (dimensions $M\,L^{-3}$), linear size (dimension L) and speed $v$ ($L\,T^{-1}$), then we might write

$$\text{Energy} = k\rho^x l^y v^z \qquad (6.1)$$

where $k$, $x$, $y$ and $z$ are unknown *dimensionless numbers*. In terms of the units in the problem,

$$[M] [L]^2 [T]^{-2} = k([M] [L]^{-3})^x[L]^y([L] [T]^{-1})^z \quad (6.2)$$

and since the units of M, L and T are fundamental, we may equate powers of [M], [L] and [T] on both sides so that

$$\left.\begin{array}{r} 1 = x \\ 2 = -3x + y - z \\ -2 = -z \end{array}\right\} \quad \text{giving} \quad \left\{\begin{array}{l} x = 1 \\ y = 3 \\ z = 2 \end{array}\right. \quad (6.3)$$

Hence we are able to derive the result that the kinetic energy scales as $L^3$, even though the constant of proportionality $k$ (actually $\frac{1}{2}$) cannot be found by this method.

In this calculation we knew the answer in advance, but there are many other cases where we do not (and in some of them we still don't know how to get an exact answer). The technique is often used, for example, to analyse fluid flows. The fluid properties that are relevant in determining the aerodynamic or hydrodynamic drag force on an obstacle (a building in the wind, or a submarine in the water) are the fluid's density $\rho$ (since the 'heavier' the fluid to be 'pushed' out of the way, the greater the drag) and its viscosity $\mu$. Viscosity is a measure of how 'sticky' a fluid is (high for treacle, low for air) and determines the frictional forces between sliding fluid layers. When a piece of paper slides across a table on a cushion of air of thickness $d$, say, the frictional force experienced by the paper is found experimentally to be directly proportional to the viscosity of the air and to the area and speed of the paper, but inversely proportional to $d$ (the thinner the cushion, the higher the frictional forces). Since force has dimensions of $M L T^{-2}$, it follows that $\mu$ has dimensions of $M L^{-1} T^{-1}$.

Consider, then, the design question of how the aerodynamic drag on a car of a given geometric shape moving at speed $v$ scales with its linear dimension L. If we suppose that at high speed the viscosity of the air is irrelevant, then the drag force can depend only on the car's length $l$, the speed $v$ and the density of the air $\rho$. Then the only possible dimensionally consistent force law of the form Drag $= k\rho^x l^y v^z$ is

$$\text{Drag} = k\rho l^2 v^2 \quad (6.4)$$

The *drag coefficient* $k$ is undetermined (and indeed is exceedingly difficult to calculate for a given car shape), but the result that the drag scales as the square of the linear dimension of the car is immediate. Car advertisements often boast that by clever design the

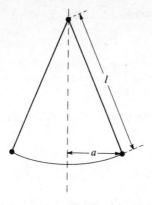

*Figure 6.2.* The simple pendulum.

car's drag coefficient has been made especially low (figures of about 0.3 are often claimed).

Second, consider the drag on a small ball-bearing dropped into a large vat of a very viscous oil. We may guess that the sphere experiences a drag force that depends only on its speed $v$ and radius $l$, and on the viscosity $\mu$ of the fluid in which it is immersed. Our scaling method shows that the only possible force law is

$$\text{Drag} = k\mu v l \qquad (6.5)$$

and the drag here depends linearly on the dimension L. (In this case a detailed calculation – involving the solution of a partial differential equation – yields $k$ as $6\pi$, a result known as Stokes's law.)

Because there are *three* fundamental mechanical units, the dimensional analysis procedure described above will, in general, be successful only when the dependent quantity (energy, say) depends on three or fewer independent variables (for energy, they are density, size and speed) for then we have three simultaneous equations for just three unknowns ($x$, $y$, $z$). Commonly, however, we do not find ourselves in that fortunate position: there are often four or more variables to take into account.

An example often glossed over in physics textbooks is the so-called 'simple pendulum' (Fig. 6.2). We would like to know how the period $P$ of the pendulum scales with its length $l$. The period can presumably depend only on the length $l$, the mass $m$ of the bob, the amplitude $a$ of the swing and the acceleration due to gravity $g$ (with dimensions $L T^{-2}$). So we write

$$P = k l^x m^y a^z g^u \qquad (6.6)$$

Then, equating powers of M, L and T, we find that

$$\left. \begin{array}{l} 0 = y \\ 0 = x + z + u \\ 1 = -2u \end{array} \right\} \quad \text{giving} \quad \left\{ \begin{array}{l} u = -\tfrac{1}{2} \\ y = 0 \\ x + z = \tfrac{1}{2} \end{array} \right. \qquad (6.7)$$

and thus $x$ and $z$ cannot be determined separately. Our scaling law is then

$$P = k \left(\frac{l}{g}\right)^{1/2} \left(\frac{a}{l}\right)^{z} \qquad (6.8)$$

Since any value of $z$ will do, we may write this result in the more general form

$$P = k \left(\frac{l}{g}\right)^{1/2} \times \text{(function of } a/l) \qquad (6.9)$$

in which the function cannot be determined by scaling methods.

At first sight it appears that our method has failed: we have been unable to disentangle completely the dependence of $P$ on $l$ because the amplitude $a$ still appears. We *can* conclude, however, that, for a given maximum angle of swing (i.e. fixed $a/l$), $P$ scales as $l^{1/2}$. Furthermore, even though our solution is not complete it still represents a major advance, in that it has reduced from three ($l$, $m$ and $a$) to one ($a/l$) the number of independent variables we need to consider ($g$ is effectively constant). Thus, as an experimenter who wishes to measure $P$, you should not seek to vary $l$, $m$ and $a$ separately, but rather should fix, say $l$ and $m$, and vary $a$. You may then measure the unknown function of $a/l$ and deduce $P$ for all other choices of $l$ and $m$. Similarly, the calculational task is much reduced: if the period can be calculated for all possible angles of swing, i.e. all possible values of the dimensionless group $a/l$, then the full solution is known. (The function of $a/l$ actually varies by only a few per cent from the fixed value $2\pi$ over the whole available range, 0 to 1. Unfortunately this mathematical serendipity does not extend to most problems, e.g. the drag calculation below.)

Scaling methods, coupled with dimensional analysis, enable us to reduce the calculational or experimental task of examining a physical phenomenon to that of investigating particular dimensionless numbers (such as $a/l$) which govern the phenomenon. For example, in the discussion of fluid drag above, we noted that *in general* the hydrodynamic force on a body may depend on its size $l$ and speed

$v$, and also on the density $\rho$ and viscosity $\mu$ of the fluid in which it is immersed. You can verify (by methods similar to those above) that this general form for the drag must be

$$\text{Drag} = \rho v^2 l^2 \times (\text{function of } \rho v l / \mu) \qquad (6.10)$$

in which the dimensionless group $\rho v l / \mu$ is called the fluid *Reynolds number*. How does this expression fit in with the special drag results given above? If the flow is fast enough, then the Reynolds number will be large, and, as in our earlier discussion, we expect the fluid viscosity to become unimportant. This occurs if and only if the function becomes constant as $\rho v l / \mu \to \infty$, and then we recover the $\rho v^2 l^2$ drag result. But if the Reynolds number is small, we expect the fluid density to be irrelevant, and this is possible only if

$$\text{function of } \rho v l / \mu \text{ varies as } (\rho v l / \mu)^{-1} \text{ as } \rho v l / \mu \to 0 \qquad (6.11)$$

and then the drag is proportional to $\mu v l$, as before.

Suppose we wish to construct a model experiment to measure the lift and drag on a novel design of aircraft wing (or yacht keel). To make a full-scale mock-up would probably be prohibitively expensive, but our result indicates the form of scaling for the drag or lift at a given Reynolds number. Thus if, say, the model wing is one-tenth real size and the fluid chosen is air in a wind-tunnel, then to maintain the same Reynolds number the flow speed needs to be ten times larger than normal. The drag should then (as it happens) be the same as for the full-scale wing. Alternatively, if the fluid used in the model experiment is water (a thousand times as dense as air, and a hundred times as viscous) then the velocity will not need to be changed to achieve the same Reynolds number, but the measured drag will be ten times larger than for the full-scale system.

## Using scaling methods to simplify calculations

We have noted that scaling methods can reduce the number of independent parameters in a physical problem by the introduction of one or more dimensionless numbers. A host of such dimensionless numbers have been introduced to help describe the more common physical phenomena. They can often provide a rational basis (though not a wholly deductive one) by which problems may be further simplified by identifying those ingredients that are unimportant.

For instance, suppose you are designing a rocket, and need to find whether the gravitational pull of the Sun will significantly affect the thrust required to enable the rocket to lift off. There is no doubt that the Sun must have *some* effect, but intuition suggests that it will be a small one. But how small? In principle you could perform two calculations, one including the Sun's pull, a second ignoring it, and compare the two. This would however be very costly and wasteful in time and effort. Much simpler is to estimate a dimensionless number that represents the ratio of the doubtful effect (the Sun's pull) to one we know to be important (say the Earth's pull). It is straightforward to verify, by means of the inverse square law for gravitation, that the ratio is about $5 \times 10^{-4}$. So, to an accuracy of about one part in a thousand, you may neglect the Sun's effect in a calculation of the rocket thrust required.

Similarly, in evaluating hydrodynamic forces we may estimate the flow Reynolds number and, if it is large, we have a rational basis for supposing that viscous effects will be negligible (giving a drag force scaling as $L^2$), whereas if the Reynolds number is small the viscous effects will dominate (and the drag will scale as L). For example, a shark is about 1 m long, and can swim through the water at about $1 \, \text{m s}^{-1}$. The density of water is $10^3 \, \text{kg m}^{-3}$, and its viscosity is $10^{-3} \, \text{kg m}^{-1} \text{s}^{-1}$. The Reynolds number for the flow past the shark is thus about $10^6$ and, as a first approximation, viscous effects may be ignored. For a swimming spermatozoon, on the other hand (length about $10 \, \mu\text{m}$, speed about $1 \, \mu\text{m s}^{-1}$), the Reynolds number is $10^{-5}$ and viscosity will win. For the flow of blood in an artery, a typical flow Reynolds number is a few hundred, whereas in a capillary it is a few hundredths, and the analysis of both flows may be made easier by means of appropriate simplifications.

The process of dimensional analysis coupled with order-of-magnitude estimation of dimensionless numbers is widely used to identify the important and the unimportant ingredients of complex physical, biological and engineering systems. A deep and subtle question which we have glossed over is whether a calculated property will have approximately the same value when a dimensionless number is small as it does when the number vanishes altogether. For instance, is the drag on a body in a slightly viscous fluid approximately the same as a drag in an altogether inviscid (zero viscosity) fluid? Surprisingly, the answer is that sometimes it is not, and the elucidation and resolution of this embarrassing complication has been and continues to be the source of a good deal of mathematical activity.

The nature of modern so-called 'asymptotic methods' lies outside the scope of a chapter on scaling, but it is instructive to see by reference to our hydrodynamic example above where our simple-minded arguments may fail. The origin of the difficulty is that in writing down a Reynolds number we tacitly suppose that it represents the relative importance of viscous effects *everywhere* in the fluid. Unfortunately this may not be the case: with high Reynolds numbers over most of the flow field viscous effects *are* negligible, but there may still be small regions (often close to the boundaries of obstacles in the flow) where viscous effects remain important, and these small regions can in their turn modify the flow elsewhere. Thus the approximation of ignoring the viscosity is non-uniform: it is good in most places but bad in others. The moral of this tale is clear: in deciding whether a physical effect may be ignored, it is certainly *necessary* for the appropriate dimensionless number to be small, but this condition may not always be *sufficient*.

## Scaling ideas in biology

It is a matter of simple observation that no present-day animal whose weight exceeds about 30 kg is able to fly. Pigs can't fly. Is this because evolutionary pressures have made flight an undesirable attribute for heavier creatures, or is there a more fundamental, physical reason for an upper weight limit? A complete answer to this question would be exceedingly complex, requiring not only a study of aerodynamics but also of factors such as muscle, bone and nervous activity, all within an exceedingly complicated anatomical geometry. Nevertheless, simple scaling ideas can offer valuable clues.

Let us first assume that all creatures of a given class (birds, say) are geometrically similar. This is plainly wrong – a sparrow does not have the same shape as an albatross. Nevertheless, provided the range of shapes is not *too* wide, we may reasonably hope that, although details of shape may change our estimates numerically, they will not affect the scaling behaviour. This is more likely to be true if we restrict the class of creatures considered (say to species of bird which adopt a particular form of flight, such as gliding or hovering). If the characteristic length scale of a creature is L, then, as noted earlier, its weight must scale as $L^3$. Now, in estimating the physical abilities of this creature (e.g. whether it should be able to fly, run or jump) we need to calculate the scaling behaviour of the muscle activity required to support this weight, so we need some biological information on the factors which limit muscle activity.

One way to proceed is to note that the steady rate at which a creature can do work to combat air resistance when flying, say, is limited by the rate at which it can take up food (or oxygen) to provide the energy input. At first sight this input rate appears to scale with the volume of the digestive system (or lungs) of the creature, and thus would scale as $L^3$. In fact, however, this *volume* is irrelevant: nutrients are absorbed by passing through the walls of the gut, and so it is the total *area* of gut wall that limits the rate of absorption. The appropriate estimate of this surface area, and thus of the rate of energy input, is therefore $L^2$.

Of course, other biological factors or hypotheses may restrict this energy output further. For instance, the metabolism of digestion generates heat. If the creature is to maintain its body temperature over a long period, then this heat must be lost to the surroundings. Now, the rate of heat loss by a creature scales with its surface area and thus, again, with $L^2$. So, apparently fortuitously, both the 'food input' and 'heat loss' hypotheses give the same scaling behaviour for the power output.

Yet a third factor which might be relevant in limiting the power output is the anatomical constraint that a given muscle strand can provide only a limited force before it breaks. Because, however, the maximum overall force that can be exerted by a muscle is proportional to the number of strands that it contains, and because this number is itself proportional to the cross-sectional area of the muscle, we find an $L^2$ scaling behaviour for the maximum force, which is again consistent with an $L^2$ scaling for the power output.

The coincidence of these three estimates provides a strong basis for supposing that the answer is correct. Additionally, however, if observation supports the conclusion that the power output *does* scale as $L^2$, any one of the three hypotheses – or indeed others – might actually be the dominant factor, and scaling ideas alone cannot distinguish between them. If, on the other hand, observations show a different scaling, all three hypotheses could be rejected at once. This is an example of the general principle that scaling ideas can provide swift and powerful tests for *rejecting* possible mechanisms, but cannot elucidate details of competing mechanisms having the same scaling behaviour.

By means of the scaling estimate for the power output, we may now infer by (in principle) exact physical laws how the athletic performance of our hypothetical class of creatures will scale. When an animal runs at its maximum steady speed $v$ on the flat, it does work against the air drag that it experiences. Provided the Reynolds number of the air motion is sufficiently high (which will certainly be the case if the animal is fairly large), it will experience a drag $k\rho l^2 v^2$, and thus must supply power at a rate $k\rho l^2 v^3$. Now, the available power supply scales as $L^2$, and so the prediction is that $v$ will be independent of $l$. Observations of, say, rabbits, dogs, humans and horses show that this result is approximately true over a wide range of sizes $l$, or at least that the variability between different species (of dog, say) having the same size, and even between different creatures within the same species, is comparable. For microscopic aquatic creatures swimming at low Reynolds numbers, the locomotive power required scales as $\mu l v^2$. So, by using the same hypothesis as for power supply, we can predict that $v$ will scale as $L^{1/2}$: larger creatures can swim comparatively faster.

The factor limiting how high an animal can jump is likely to be the maximum force that can be supplied by muscles rather than the steady output of power. As noted above, however, this also scales as $L^2$, so since the available muscular extension is also anatomically determined (by the length of the thigh, say), the work done in flexing the muscles required for jumping scales as force × distance, proportional to $L^3$. If the animal jumps to a height $h$, this work is converted into potential energy $mgh$, where $m$ is the mass of the creature and $g$ the acceleration due to gravity. Now, since $m$ scales as $L^3$, $h$ is independent of L, so we can predict that, broadly, small jumping creatures can jump as high as big ones. Again this relationship is borne out over a wide range of length scales. A great grey kangaroo, about 1.6 m tall, can jump to a height of about 1.5 m. Jerboas leap in much the same fashion as kangaroos, and despite being only a tenth as tall they can jump to a height of over 1 m.

The scaling analysis of animal flight is more difficult, in particular because of the extreme aerofoil shapes of birds' wings. But a simplified treatment can show us that steady flight by large creatures is impossible. Suppose a bird flies steadily and horizontally. Then its weight, which is proportional to $L^3$, is supported by an aerodynamic lift force, which must also scale as $L^3$. Since, however, the lift scales as $\rho l^2 v^2$, the speed of flight $v$ must be proportional to $L^{1/2}$. Now, the drag force, though numerically smaller than the lift, has the same scaling behaviour, so the power required to maintain a steady speed scales as drag × speed, or $L^{7/2}$. The power available for flight scales as $L^2$ and, as L increases, this rises more slowly than $L^{7/2}$. Hence there must exist a maximum value for L, and hence also for mass, beyond which steady horizontal flight is unsustainable. (A more careful and difficult analysis of the flow over a wing of high aspect ratio (one having *two* very different L scales, length and breadth) may slightly reduce the 7/2 exponent of L, but not below 2. So the result that there must be an upper limit on mass remains valid.) Our scaling analysis cannot predict what the maximum mass will be. Observation suggests a figure of about 30 kg.

The remarkable feature of these biological examples is their ability to provide 'something for nothing'. Very crude estimates, without detailed mechanisms, enable remarkably strong predictions to be made, albeit of a qualitative nature. In many instances, because there is, inevitably, variation within biological populations, crude estimates are the best we can do; indeed, anything more refined would not be justified.

### Fractals

We have seen that, while linear dimensions scale as L, surfaces scale as $L^2$ and volumes as $L^3$. We also saw that physical properties, such as the period of a pendulum, may scale as some *fractional* power of L. A natural (pure) mathematical question is whether geometrical objects might exist whose dimension is intermediate between 1 and 2: an object 'denser' than a line, but 'less dense' than a surface. Surprisingly, perhaps, such objects of fractional dimension (called *fractals*) can be constructed; more surprising still, they are currently the focus of a good deal of *physical* attention.

A simple mathematical example of a fractal is the *Sierpinski gasket*, shown in Fig. 6.3. It may be constructed by the following sequence of steps. From an equilateral triangle ABC the central triangle B'C'A', formed by joining the bisectors of its sides, is removed. Three smaller triangles (e.g. AB'C') are left, and each of these in turn has its central triangle removed to give nine yet smaller triangles. When this process is carried to infinity a very 'lacy' structure is left, filled with infinitely many holes. The structure that remains is *self-similar* in the sense that under a microscope the triangle AB'C' looks exactly the same as the original figure ABC, and this structural similarity recurs at smaller and smaller length scales.

Supposing it were possible to make such a gasket out of sheet metal with uniform thickness and density. How would its weight scale with L? The weight is proportional to the area of the gasket, so if there were no holes it would be proportional to the area of the triangle and hence scale with $L^2$. With the holes, the weight w of ABC is just three times the weight of AB'C'. If, then, we tentatively assume that

$$w = kl^d \tag{6.12}$$

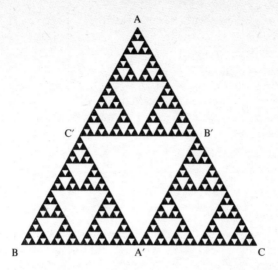

*Figure 6.3.* The Sierpinski gasket. From the middle of a triangle ABC the triangle A′B′C′ is removed. The process is repeated with each of the three remaining triangles, then with each of the nine left after that, and so on *ad infinitum.* (Here only the first six generations of the gasket are shown.) The result is a 1.58-dimensional object!

for some constant $d$, the 'fractional dimension' of the gasket (of side $l$), then, since AB′C′ is a similar triangle of side $l/2$, we must have

$$kl^d = 3k(l/2)^d, \quad \text{giving} \quad 2^d = 3$$

and so

$$d = \ln 3/\ln 2 = 1.58\ldots$$

Similar hole-boring operations enable us to construct objects with dimension between 0 and 1 (by chopping portions out of a line) or between 2 and 3 (by drilling holes in a cheese).

Weird structures like these cannot be characterized by a single length scale L: they have structure at *all* length scales smaller than L. Do such apparently abstract constructs have any parallel in nature?

Consider the at first sight reasonable task of measuring the length of the coastline of the UK. One can imagine a diligent geographer measuring a $1:10\,000$ Ordnance Survey map with a ruler. Unfortunately, however, this procedure would not produce the 'true' answer because small bays and inlets would not be apparent on such a map, and a finer-resolution map would almost certainly give a

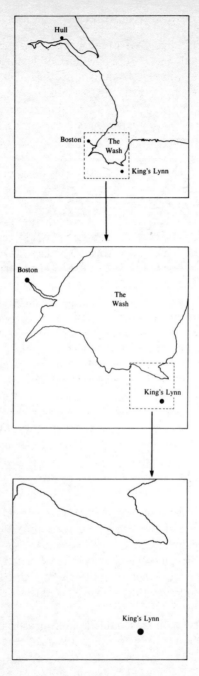

*Figure 6.4.* How long is the coast of Britain? As the resolution increases, so does the amount of detail visible – and so, apparently, does the total length of coastline.

much larger result (see Fig. 6.4). An even more conscientious observer on the seashore would find even tinier coastal features, such as individual rocks, and ultimately individual grains of sand, and would arrive at an even larger estimate. In summary, the answer will depend on the chosen length scale of resolution, and it is this presence of structure on many scales (here from sand grain dimensions up to coastal dimensions) that is qualitatively characteristic of fractal structures.

Fractals have recently been seen quantitatively (and fractal dimensions measured) for a number of *aggregation* processes. Small particles suspended in a fluid are continually jiggled by Brownian motion (the constant random motion of individual molecules), and occasionally two will bump into one another. If the forces of attraction between the two particles are sufficient, they may then stick together to form an aggregate which then collects further particles as time proceeds, each sticking to what is at that moment the outer surface of the aggregate. This process of *diffusion-limited aggregation* characteristically results in the formation of large open clusters, full of holes, with a measured fractal dimension (over different scales from the size of a single particle upward) of about 1.7.

The same process of fractal generation can be simulated on a microcomputer. A 'particle' is released at random some distance from an 'aggregate', and then takes a series of random steps. If it bumps into the aggregate it sticks, and increases the size of the aggregate, and a new particle is released. After many such random releases the aggregate takes on a fractal appearance. A simulation of this sort on a large computer has aggregated over four million such particles, and the fractal dimension of the resulting aggregate was indeed close to 1.7.

Similar fractal structures have been seen in such diverse phenomena as lightning bolts that fork repeatedly, electroplating and polymers. An economically important case of interest is oil recovery. Oil is often trapped in the interstices within porous rocks. The geometry of the pore spaces is exceedingly complex, but it too conforms to a fractal structure over a wide range of length scales. The task of examining the dynamics of the trapped oil, and thereby comparing the relative efficiencies of different oil recovery strategies, is the focus of a good deal of current research.

## Conclusion

Confronted with a new or unfamiliar problem to analyse, the prudent engineer or scientist will seek to identify the important physical ingredients that need to be incorporated in any mathematical model. He or she will not wish to embark on an analysis of full complexity at the outset, preferring instead to estimate the effectiveness of competing mechanisms on the basis of a 'back of the envelope' calculation. We have seen how the process of investigating scaling behaviour, coupled with dimensional analysis and order-of-magnitude estimation, provides just the sort of 'guesstimate' required – and the complexity of calculation and the lack of data often mean that in practice a more accurate analysis would not be justified.

## Further reading

J. Maynard Smith, *Mathematical Ideas in Biology* (Cambridge University Press, 1968).

T. J. Pedley (Ed.), *Scale Effects in Animal Locomotion* (Academic Press, 1977).

L. M. Sander, 'Fractal growth', in *Scientific American*, Jan. 1987.

# 7

## BIOLOGICAL MODELLING: POPULATION DYNAMICS

### JOHN MACQUEEN

#### Working models

More and more, mathematics is being applied in medicine, biology and ecology. It might be argued that the degree of formally disciplined thought required in various branches of science can be described by a hierarchy, descending in the order mathematical, physical, chemical and biological. On the other hand, the complexity of the systems which are studied in these branches can be described by the same ordered list, but in ascending order.

Mathematical modelling is the process of idealizing systems in the real world, by abstracting a number of mechanisms considered to be significant and describing them using the formalism and precision of mathematical methods. It is really only in the last hundred years or so, since Thomas Malthus considered the problem of human population growth, that mathematical ideas have been applied directly to living systems. Nowadays one finds mathematical methods being used to describe human blood flow and neurone activity, vision and hearing, and genetics, as well as the transport of nutrients in ecological systems and pathways for pollutants to living species through water, land and air, and by food intake and breathing. These are complicated systems, governed by many parameters such as climate, geology and even social customs, which are clearly difficult to describe in mathematical terms. Nevertheless, the challenge which living systems pose to the rigour of mathematics (pure and applied) is irresistible.

In this chaper we shall see how fairly simple differential equations – equations containing derivatives as defined in differential calculus – may be used to describe certain aspects of growth and control in biological populations, and to highlight the interesting biological and mathematical properties of these equations. From our starting point, analysis, we shall go on to look at non-linearity, chaos and numerical simulation.

A nice class of problems is provided for mathematicians by the general requirement to model the growth and regulation of fish, animal or crop species used by mankind as food sources. For example, the numbers of herring in the North Sea, tuna in the eastern tropical Pacific and baleen whale in the Antarctic are all affected by governmental policies on fishing in these areas. There are two obvious long-term consequences of over-fishing: reduced catches of the species in question, and a disturbance of the ecological balance – for example by allowing the growth of another species or the decay of a predator – with perhaps unforeseen consequences. Thus mathematics has a role to play in the formation of husbandry policies, and may shed light on the interpretation of historical data on fishing, leading to changes of policy.

We should not become too enthusiastic about this type of application. The assumptions behind a mathematical model may be reasonable, but the model is bound to be very idealized, and it may even be quite wrong. The data used to define the problem parameters, such as growth or population density, may be of poor quality and so the significance of the decimal places should not be overestimated. However, mathematics in population dynamics has its place, as we shall now show.

### Mathematical building blocks

First, though, we look at some of the mathematics behind the mathematical modelling: exponential and logarithmic functions and their derivatives, and the method of partial fractions.

We begin with the *exponential function*, defined to be a function $f(x)$ (or simply $f$) of the variable $x$ with the property that the slope of its graph at any point $x$ is equal to the value of the function at that point. This means that if the function $f(x)$ exists (which we really ought to prove, but will boldly assume for now) then it is related to its derivative with respect to $x$ by

$$\frac{df}{dx} = f \tag{7.1}$$

Here we have an example of a *differential equation* – an equation containing differential coefficients – for the function $f(x)$. Here $f$ can be any multiple of the function $e^x$ (or $\exp(x)$), referred to as the

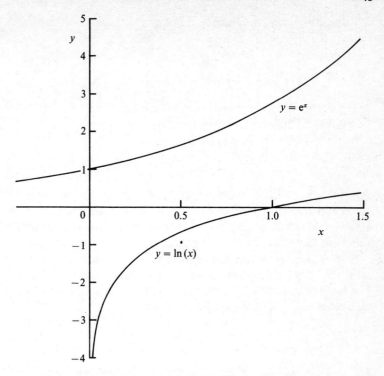

*Figure 7.1.* The exponential function $y = e^x$ and the logarithmic function $y = \ln(x)$.

exponential function, where

$$e = 2.781\,28\ldots = \lim_{n \to \infty}\left(1 + \frac{1}{n}\right)^n \tag{7.2}$$

The history of e is intimately linked with the study of financial rather than biological growth – with the calculation of compound interest.

In the same way that taking a square root is the inverse of squaring, so the *logarithmic function* – which is written as $\ln(x)$ – is the inverse of the exponential. Thus we write

$$y = \ln(x) \Leftrightarrow x = \exp(y) \tag{7.3}$$

The graphs in Fig. 7.1 show how the functions $\exp(x)$ and $\ln(x)$ vary over a range of values of $x$. You can see that $\exp(x)$ is always positive, so that $\ln(x)$ is not defined for negative values of $x$.

From this definition and the differential equation for $\exp(x)$, it follows that if $y = \ln(x)$ then

$$\frac{dy}{dx} = \frac{1}{x} \tag{7.4}$$

This differential equation can be rearranged into a differential form which will arise several times during the forthcoming biological discussion:

$$dy = \frac{dx}{x} \tag{7.5}$$

When mathematicians come across this expression they recognize it as an old friend: it integrates to $y = \ln(x) + C$, where $C$ is an arbitrary constant (corresponding to the arbitrary multiple of $\exp(x)$ which we have already encountered).

Now we turn from calculus to algebra, and introduce a technique for simplifying certain expressions by the method of *partial fractions*. The biological models with which we shall be concerned sometimes require us to evaluate expressions similar to

$$\int \frac{dx}{(x + 2)(x + 3)} \tag{7.6}$$

The major complication here is the form of the denominator – which the method of partial fractions provides a way of simplifying. The aim is to calculate constants $A$ and $B$ which, for all values of $x$, satisfy the identity

$$\frac{1}{(x + 2)(x + 3)} \equiv \frac{A}{x + 2} + \frac{B}{x + 3} \tag{7.7}$$

If $A$ and $B$ exist, then we must have

$$\frac{1}{(x + 2)(x + 3)} \equiv \frac{A(x + 3) + B(x + 2)}{(x + 2)(x + 3)} \tag{7.8}$$

and hence the numerators must be identical:

$$1 \equiv (A + B)x + (3A + 2B) \tag{7.9}$$

This identity can be maintained only if the coefficients of $x$ and the constant terms on both sides are identical, which requires that

$$3A + 2B = 1, \qquad A + B = 0 \tag{7.10}$$

The constants $A$ and $B$ are determined from this pair of simultaneous equations to be $A = 1$ and $B = -1$. Hence our original integral is simplified to

$$\int \frac{dx}{(x + 2)(x + 3)} = \int \frac{dx}{x + 2} - \int \frac{dx}{x + 3}$$

$$= \ln(x + 2) - \ln(x + 3) + C \quad (7.11)$$

where $C$ is the arbitrary constant of integration.

This digression into calculus and algebra should make clearer some of the mathematical manipulations of biological equations that follow.

## Simple ideas about growth

The use of mathematics to describe the growth and decay of living organisms interacting in a biological or ecological system requires some familiarity with the exponential function $e^t$ and its associated differential equation to represent a quantity which grows with time $t$. So first we look at a few ideas about growth and its description in mathematical terms.

Consider, for example, the distance covered by a train travelling at 50 mph from Glasgow to Manchester. In 1 hour it will have gone a distance of 50 miles, in 2 hours this will have increased to 100 miles, and so on – the distance $s$ of the train from Glasgow increases steadily with the passage of time $t$ (in hours) according to the formula

$$s = 50t \quad (7.12)$$

In more general terms, if the train's speed is $u$ then the distance $s$ depends on travel time $t$ according to the relation

$$s = ut \quad (7.13)$$

Such a relation describes the simplest kind of growth, in which the rate of increase is constant. The growth is then described as linear in time (because the graph of distance against time is a straight line, as we have seen in Chapters 1 and 6).

Although it may seem a little heavy-handed, it will suit our purpose to introduce here a little differential calculus. The velocity $u$, being the rate of change of distance with time, can be calculated as the derivative of $s$ with respect to time $t$:

$$ds/dt = u \quad (7.14)$$

This is a *differential equation* – an equation involving derivatives – which determines the rate of growth of distance with time.

A different type of growth is found when one considers the volume of a cube as a function of the length of its side. A box with side 1 m has a volume of $1 \, \text{m}^3$; doubling the side to 2 m increases the volume to $8 \, \text{m}^3$ and for a 3 m cube the volume is $27 \, \text{m}^3$. In this case, the growth of the volume $V$ is related to the side length $l$ by

$$V = l^3 \tag{7.15}$$

This non-linear formula determines the volume as a function of the side length. It can be seen that increasing the side from 1 m to 2 m increases the volume by $7 \, \text{m}^3$ (from $1 \, \text{m}^3$ to $8 \, \text{m}^3$); increasing the side from 2 m to 3 m increases the volume by $19 \, \text{m}^3$. Equal increases in the length do *not* yield equal increases in volume. This behaviour is a feature of non-linear relationships, and corresponds to the mathematical statement that the rate of change of volume with length of side depends on the length of side. The phrase 'rate of change' suggests that we can use differential calculus here. We know that if $V$ and $l$ are related by equation (7.15), then

$$dV/dl = 3l^2 \tag{7.16}$$

This is a differential equation that determines the rate of growth of volume as a function of side length $l$.

We come now to a somewhat more complicated kind of growth, but one closer to the population dynamics we shall be discussing. Consider the sequence of numbers formed by starting with unity and doubling each time. These numbers form the sequence

$$S = 1, 2, 4, 8, \ldots, 2^n, \ldots \tag{7.17}$$

Similarly, successive increase of each number by 50% generates the sequence $1, 1.5, 2.25, \ldots, (1.5)^n, \ldots$. In general, if $a$ is any positive number then the sequence $a^n$ can be calculated and the numbers within it are said to grow *exponentially*. Consideration of such numbers leads us to generalize a little and discuss the features of a function $y(x)$ defined by

$$y = a^x \tag{7.18}$$

where $a$ is a real number greater than one, and $x$ is real – positive or negative. A little experimenting with calculator, pencil and graph-paper will show that $y(x)$ increases without limit as $x$ increases positively, and decreases towards zero (becoming vanishingly small) as $x$ decreases through negative values. The feature of this type of growth which I want to stress is that the rate of growth of $y$ is proportional to $y$: the increase in $y$ for unit increase

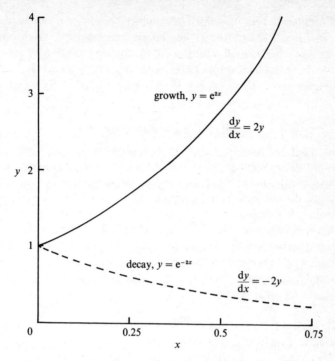

*Figure 7.2.* Exponential growth and decay.

in $x$ is $(a - 1)y(x)$. The mathematical equation that describes such growth – one in which the rate of change of a quantity is proportional to the quantity itself – is

$$dy/dx = ky(x) \qquad (7.19)$$

where $k$ is a constant. The general solution of this equation is known to be

$$y(x) = A e^{kx} \qquad (7.20)$$

where $A$ is an arbitrary constant; e we have met in equation (7.2). For example, it is easy to see that if $k = 2$ and $y(x) = A e^{2x}$ then $dy/dx = 2Ae^{2x} = 2y(x)$. The value of $A$ is determined by specifying the value of $y$ for some (single) specific value of $x$; usually we wish to start with a value of $y$ when $x = 0$, corresponding to an *initial condition* which fixes the constant $A$. The differential equation in this case is said to describe exponential growth if $k > 0$; when $k < 0$ the solution $y$ decays steadily towards extinction (zero). This behaviour is shown in Fig. 7.2, with $k = \pm 2$ and $A = 1$.

To summarize, we have discussed a variety of examples of growth, from linear to exponential systems, each of which has an associated differential equation. It will become clear that the concept of exponential growth has particular relevance to population dynamics.

## Single-species population dynamics

The most commonly discussed mathematical models of a single biological population are described by a differential equation which specifies the rate of change (growth) of the population $P(t)$ for a known growth rate $R(P)$ as follows:

$$dP/dt = R(P) \tag{7.21}$$

This equation is to be solved for $P$ as a function of time $t$, subject to an initial condition of the form $P = P_0$ when $t = 0$. A particularly simple model of this kind is the so-called Malthusian or exponential growth described by

$$dP/dt = kP \tag{7.22}$$

where $k$ is the *intrinsic growth rate* per unit of the population. The solution of this linear differential equation is $P = P_0 e^{kt}$, which grows without limit or decays to extinction depending on whether $k$ is positive or negative, as shown in Fig. 7.2. This form of $P(t)$ grows at a rate proportional to its current size, just as a sum of money does when invested in a bank or building society.

The intrinsic growth rate $k$ is a parameter in the rather idealized biological growth system which equation (7.22) represents. We can think of it as defining the rate at which the population of a species might grow in the absence of any competition for or limitations on food. For example, if a small number of organisms are introduced to a large and suitable environment, they will tend to reproduce rapidly in line with the Malthusian rate. A remarkable example of this is known to have occurred with a species of fish, the striped bass. In about 1880, 435 of these fish were released into San Francisco Bay. Just twenty years later the commercial catch alone exceeded 550 tonnes – which suggests at least a millionfold increase over that period. A similarly explosive growth in the human species was foreseen by Malthus in the last century. Such an unrestricted increase of numbers in a population, however, is not sustained in practice because of limitations imposed by several factors, one of which is the finite food supply.

For many species living in a stable environment with a finite food supply, the population tends to a stable upper limit which depends on the environmental conditions. Below this limit, the *environmental carrying capacity* or *saturation level*, which we denote by $L$ (positive), the population has an inherent or potential rate of growth which depends on the population size. Biological species, from micro-organisms to mankind, are believed to be growth-limited in this fashion. A particular form of equation (7.21) matching such behaviour is a refinement of equation (7.22):

$$\frac{dP}{dt} = kP\left(1 - \frac{P}{L}\right) \qquad (7.23)$$

where $k$ ($> 0$) is the intrinsic or potential growth rate, achieved by the species for as long as its population is not large enough to strain the available food supply.

As a particular example of this general equation we take $k = 2$ and $L = 1000$. Then

$$\frac{dP}{dt} = 2P\left(1 - \frac{P}{1000}\right) \qquad (7.24)$$

This is a mathematical way of stating that at any particular moment the growth rate is 2 units per unit of population (say 1000 fish) per unit of time (a week say), but that in addition the fish population is limited to a maximum of 1000 by environmental constraints such as the availability of food. You can see that if the population $P$ in equation (7.24) were to be greater than 1000, then the rate of growth would be negative and the population would decrease from such a value. This would happen if the owner of a fish farm overstocked his ponds in a misguided attempt to rear more fish.

The differential equation (7.24) can be solved by the method of partial fractions. To calculate the complete solution we must know how many fish there were to start with. If we assume there to have been 20, then it turns out that at a later time

$$P = \frac{1000}{1 + 49e^{-2t}} \qquad (7.25)$$

Notice that this equation satisfies the condition $P = 20$ when $t = 0$. It also indicates that $P$ approaches 1000 for large values of $t$, without ever actually reaching it. A graph of $P$ as a function of time $t$ is shown in Fig. 7.3. The greatest growth rate occurs when the graph is steepest, which is when $dP/dt$ has its maximum value. From

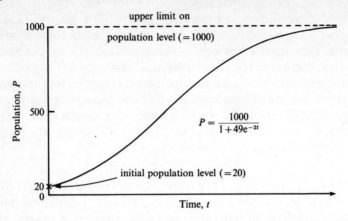

*Figure 7.3.* Limited population growth.

equation (7.24) it follows that this will be when $P[1 - (P/1000)]$ is greatest; this is when $P = 500$.

There are two other starting values worth mentioning, because they give rise to steady population levels, unvarying with time. First, as is obvious, if the initial value of the population is zero then it will be zero for all later times; second, if it is initially 1000, then it will remain constant at this value. These unvarying solutions are called *equilibrium states*, and it is of interest to see what happens if, rather like prodding a tower of bricks, we disturb them a little to examine their *stability*.

First we examine the stability of the solution $P = 1000$. Suppose we perturb the population, by adding a small number $p$ of population units to the existing 1000 units, so that $P = 1000 + p$. Then, from equation (7.24) we have

$$\frac{dp}{dt} = \frac{-2(1000 + p)p}{1000} \approx -2p \qquad (7.26)$$

The approximation is justified because $p^2/500$ is very much smaller than $p$. This shows that $p$ is proportional to $e^{-2t}$, and so $p$ will decay to nothing with the passage of time $t$; in other words, a slight disturbance to the population $P = 1000$ dies away and the level eventually returns to 1000. This stability therefore corresponds to that of a stable tower of bricks which sways slightly after being prodded, but soon returns to the vertical.

The equilibrium of the state $P = 0$, however, is unstable. If $P$ is small, say $P = p$, then equation (7.24) approximates to

$$\frac{dp}{dt} = 2p \qquad (7.27)$$

and so $p$ is proportional to $e^{2t}$, which increases with time. Thus the disturbance grows in a way that corresponds to our tower of bricks swaying so much after being prodded that it falls over. The biological significance of this is that if we could exterminate this population completely it would remain extinct, but if we overlooked just a tiny population, capable of breeding, then the population would grow rapidly and re-establish the non-zero equilibrium state.

## The influence of harvesting

In equations (7.23) and (7.24) the population $P$ is influenced by its intrinsic growth rate $k$ and the carrying capacity $L$ (the maximum population size the environment can support). One reason for our mathematical modelling of population growth is to get a better picture of the effect of man's predation or harvesting of a particular species, for example by fishing for herring. It is quite likely that the amount of harvesting through fishing will be proportional to the fish population since, for example, if a trawler shoots its nets into densely populated waters the catch will be greater than in sparse waters. We can modify (7.23) to include the effect of such predation by writing it as

$$\frac{dP}{dt} = kP\left(1 - \frac{P}{L}\right) - FP \qquad (7.28)$$

where the constant $F$ is the harvesting (fishing or culling) effort per unit of the population. With our previous figures of $k = 2$ and $L = 1000$, we have

$$\frac{dP}{dt} = (2 - F)P - \frac{2P^2}{1000} = P\left[(2 - F) - \frac{2P}{1000}\right] \qquad (7.29)$$

which can be compared to equation (7.24) with the intrinsic growth rate reduced to $2 - F$. Clearly, if $F > 2$ the population will eventually reduce to zero, and the species will become extinct through over-harvesting. However, if $F < 2$ then $P$ will have a stable value when $dP/dt = 0$. This value occurs with $P = P_s$, where $P_s$, the population level stable in the face of harvesting, satisfies the equation

$$2 - F - \frac{2P_s}{1000} = 0 \qquad (7.30)$$

that is, when

$$P_s = 500(2 - F) \qquad (7.31)$$

Compare this with the stable value of 1000 before we considered harvesting.

The harvested yield $Y$ obtained from the population $P_s$ is $FP_s$, and so

$$Y = 500F(2 - F) \tag{7.32}$$

Thus $Y$ is the sustainable yield, or fishing catch, corresponding to the fishing effort $F$. Obviously, as $F$ increases then so does the size of the catch: if $F$ is close to but less than 2, or close to but greater than zero, the yield $Y$ is small. You can verify graphically or by using differential calculus that the largest yield will be obtained by adjusting fishing strategy and techniques so that $F = 1$, in which case the catch will be 500.

More generally, the optimum harvesting rate is $k/2$, where $k$ is the intrinsic growth rate of the harvested species and the corresponding maximum sustainable yield is $kL/4$, where $L$ is the environmental carrying capacity; any greater or lesser effort will lower the yield.

Of course, idealized analyses of the kind outlined here must not be taken too seriously by ecological planners. So many features of the real environment have been ignored – including interactions with other species, and random fluctuations of the environment and in the distribution of species. Nevertheless, this example shows how potentially useful concepts can emerge from very simple models. Our results suggest that in order to determine the optimum harvesting effort, the intrinsic growth rate and the environmental carrying capacity of the target population should be found. These might form the basis of recommendations to a scientist or civil servant who is considering the advisability of allowing commercial fishing in a particular sea area.

### A discrete approximation suggesting chaos

We approached our discussion of population growth, based on equation (7.23), by using an exact mathematical analysis. This was possible only because our model was a highly idealized one, and therefore highly simplified. Of course, real systems are rather complicated and not amenable to such a precise analysis, so only approximate (though very accurate) solutions can be obtained, by numerical methods designed for use with a calculator or computer. For this reason, and because of the interesting mathematical behaviour arising in the process of approximation, we shall now look at numerical (or discrete) approximations to equation (7.23).

There are many biological species, such as fruit flies and other crop pests, for which the population of one season lays eggs and then dies out; during the next season the eggs hatch and the population matures, and lays *its* eggs. The generations succeed one another season by season. Such a population, with non-overlapping generations, is well suited to study by ecologists. If we number the seasons 1, 2, . . . , $n$, . . . , the population in the $n$th season can be denoted by $P_n$. As the population in one season is determined entirely by the population of the preceding season, then, in mathematical terms, $P_{n+1}$ is a function of $P_n$.

In these seasonal populations there is a tendency for the population to increase from one generation to the next when it is small and to decrease when it is large; this instinctive form of population control is a response to a limited food supply. A mathematical expression for the dependence of $P_{n+1}$ on $P_n$ in this case is

$$P_{n+1} = 4\lambda P_n(1 - P_n) \tag{7.33}$$

A curve of the form $y = 4\lambda x(1 - x)$ is called a *parabola*, and equation (7.33) states that the sequence of numbers $P_n$ it defines lie on such a curve.

We might well wonder what the numbers* $P_n$ actually look like, and it is not very difficult to investigate specific examples. For instance, suppose that $\lambda = \frac{1}{2}$ and start from $n = 0$ with $P(0) = \frac{1}{4}$. Then, from equation (7.33), we can generate the sequence of numbers $P_0, P_1, P_2, \ldots$ as follows:

$$P_0 = \frac{1}{4} = 0.25$$

$$P_1 = 4 \times \frac{1}{2} \times \frac{1}{4}\left(1 - \frac{1}{4}\right) = \frac{3}{8} = 0.375$$

$$P_2 = 4 \times \frac{1}{2} \times \frac{3}{8}\left(1 - \frac{3}{8}\right) = \frac{15}{32} = 0.469 \tag{7.34}$$

$$P_3 = 4 \times \frac{1}{2} \times \frac{15}{32}\left(1 - \frac{15}{32}\right) = \frac{255}{512} = 0.498$$

$$P_4 = 4 \times \frac{1}{2} \times \frac{255}{512}\left(1 - \frac{255}{512}\right) = \frac{65\,535}{131\,072} = 0.500$$

You can see that this sequence rapidly approaches a value (0.5) which remains unchanging from season to season (or generation to

* Here $P_n$ is a fraction of some notional maximum population.

generation). As mathematicians we are inclined to ask why this happens in this instance, and whether it happens generally with any sequences generated by equation (7.33). The answer to this general question is 'No', but to discuss the topic further we need to go through some mathematical analysis.

First we note that equation (7.33), which defines $P_{n+1}$ in terms of a growth rate $\lambda$ and the population level $P_n$ of the preceding generation, possesses equilibrium points where $P_n$ does not change from one season to the next. In these circumstances $P_{n+1} = P_n$, and so this equilibrium value $P$ satisfies a quadratic equation $P = 4\lambda P(1 - P)$; it follows that $P = 0$ or $P = 1 - 1/(4\lambda)$.

Suppose that $P$ is perturbed slightly from the equilibrium at $P = 0$, say to $p$, where $p$ is small. Then an approximate form of equation (7.33), using the small steps $p_n$ but ignoring the second power of $p_n$, demonstrates that $p_{n+1} = 4\lambda p_n$. In other words, if $0 < 4\lambda < 1$, the perturbation decays away and $P = 0$ is a stable point, or *attractor*, of the system. If $4\lambda > 1$, this disturbance grows, the equilibrium is unstable and $P = 0$ is a called a *repellor*.

Similar analysis of the equilibrium at $P = 1 - 1/(4\lambda)$, perturbing by a small $p_n$, yields, to first order (i.e. ignoring powers of $p_n$), the relation

$$p_{n+1} = 2(1 - 2\lambda)p_n \qquad (7.35)$$

If the successive perturbation steps $p_n$ are to decrease in absolute magnitude – which they will for a stable equilibrium in which the perturbation decays and $P$ returns to its starting point – we require that $-1 < 2(1 - 2\lambda) < 1$, from which it follows that

$$\tfrac{1}{4} < \lambda < \tfrac{3}{4} \qquad (7.36)$$

For these values of $\lambda$ the equilibrium is stable and $P = 1 - 1/(4\lambda)$ is an attractor; for all other values of $\lambda$ the equilibrium is locally unstable. The behaviour at successive steps can be illustrated graphically on a microcomputer; results obtained in this way are shown in Fig. 7.4, for $\lambda = 0.125$ and $\lambda = 0.725$.

We can also show from equation (7.33) that, if $0 < P_n < 1$ and $0 < \lambda < 1$, then $0 < P_{n+1} < 1$; the maximum occurs if $P_n = \tfrac{1}{2}$ when $P_{n+1} = \lambda \ (< 1)$. There is something interesting here because, although we would expect local stability only if $0 < \lambda < \tfrac{3}{4}$, this last conclusion has proved that the solutions $P_n$ will remain bounded (though presumably cycling in some fashion) if $\tfrac{3}{4} < \lambda < 1$.

In view of this, it is tempting to investigate the behaviour of the sequence of $P_n$ steps defined by equation (7.33) as $\lambda$ increases from

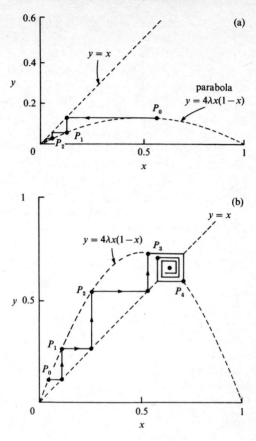

*Figure 7.4*. An attractor and a repellor for equation (7.33),
$P_{n+1} = 4\lambda P_n(1 - P_n)$, for $0 < \lambda < 0.75$. (a) The origin is an
attractor for $0 < \lambda < 0.25$. The locus of $P$ is shown here for
$\lambda = 0.125$. (b) The origin is a repellor and $1 - 1/(4\lambda)$ an
attractor for $0.25 < \lambda < 0.75$. The locus of $P$ is shown here
for $\lambda = 0.725$.

zero through $\frac{3}{4}$ to unity. We can simplify matters by concentrating on
the behaviour of the equilibrium points at $P = 0$ and $P = 1 - 1/(4\lambda)$
as the growth rate $\lambda$ is varied from zero to unity.

The transition from stable (attractor) to non-stable behaviour can
be demonstrated easily using computer graphics. For values of $\lambda$ less
than $\frac{3}{4}$, the population tends to either zero or $P = 1 - 1/(4\lambda)$, as
illustrated in Fig. 7.4. When the growth rate $\lambda$ is increased above
$\frac{3}{4}$, the population level no longer settles to a steady state. Instead

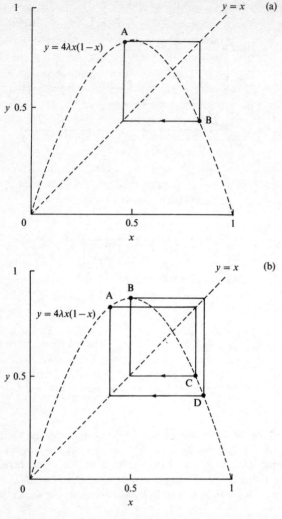

*Figure 7.5.* Bifurcation from equation (7.33), for
$0.75 < \lambda < 1$. (a) At $\lambda = 0.85$ there are two cycle points, A
and B. (b) At $\lambda = 0.87$ there are four cycle points, A, B, C
and D.

it oscillates between a sequence of, at first, two values, and then,
when $\lambda > 0.8624$, four values – the population cycles through these
values in successive seasons, as illustrated in Fig. 7.5. This splitting,
or *bifurcation*, of the equilibrium continues through sequences of
8, 16, . . . , $2^n$, . . . points.

Beyond a critical value just above 0.89, the structure of the sequence becomes apparently irregular, and this region, where $0.892 < \lambda < 1$, is described as *chaotic*. The critical value of $\lambda$ is the limit of an infinite sequence of doublings with an infinity of attractors now associated with the cycling. The system behaves as no ordinary attractor – it is an example of what is called a *strange attractor*. You may find it surprising that a stepwise procedure, or *iteration*, based on a curve as simple as the parabola can give rise to such a rich variety and complexity of behaviour. It is fascinating to watch this complicated behaviour being generated graphically on a computer. All you need are a few lines of code to generate the steps according to equation (7.33), and the simplest of graphics showing the plot of pairs $(P_n, P_{n+1})$ on your screen for various values of $\lambda$.

### Simple predator–prey systems

A limitation of the models we have considered so far is that they deal with only a single species, and ignore competitive interaction with other species. Now we look at the modelling of multi-specied populations by considering systems in which one species is preyed on by another.

Consider an ecological system in which there are *hunters* who feed on *prey*, who themselves have an unlimited source of food; as an instance we might think of birds and earthworms. A simple but illustrative mathematical model can be constructed if we make a few (sweeping) assumptions:

(1) If there were no hunters, the prey would grow without limit.
(2) If there were no prey the hunters would eventually die out.
(3) The rate at which the prey are captured is proportional to the product of the prey and the hunters – so that if the number of hunters or of prey increases, then so does the number of captured prey.
(4) The rate at which the hunters increase in number is proportional to the product of the hunter and prey populations – so that if either the number of hunters or the availability of prey increases, then so does the number of hunter births.

A biological system with these properties can be represented by the following pair of equations, to be solved for the populations $H$ and $P$:

$$\frac{dP}{dt} = (r - aH)P, \qquad \frac{dH}{dt} = (bP - s)H \qquad (7.37)$$

Here the prey has an intrinsic *growth rate*, *r*, which would be achieved if there were no hunters, while the hunters would die away with an intrinsic *death rate*, *s*, in the absence of prey. The terms involving products of $H$ and $P$, $-aHP$ and $bHP$, define how the presence of hunters inhibits the growth of prey, while the presence of prey encourages the growth of hunters.

To make equations (7.37) more specific, suppose that $a = 2$, $b = 1, r = 100$ and $s = 1000$. Then

$$\frac{dP}{dt} = (100 - 2H)P, \qquad \frac{dH}{dt} = (P - 1000)H \qquad (7.38)$$

There are two equilibrium states for which $dP/dt = dH/dt = 0$. You can check, if necessary by substituting the following values for $P$ and $H$ into equations (7.38), that these states are either

$$P = 1000 \quad \text{and} \quad H = 50 \qquad (7.39)$$

or

$$P = 0 \quad \text{and} \quad H = 0 \qquad (7.40)$$

Suppose that the equilibrium (7.39) is slightly disturbed so that $P = 1000 + p$ and $H = 50 + h$, where $p$ and $h$ are small. Then, approximately,

$$\frac{dp}{dt} = -2000h, \qquad \frac{dh}{dt} = 50p \qquad (7.41)$$

By taking the ratio of these equations, the variable $t$ (time) drops out and we obtain the following relation between the small disturbances $p$ and $h$:

$$\frac{dp}{dh} = -\frac{40h}{p} \qquad (7.42)$$

This can be solved by the techniques of separating the derivative, giving

$$\int p\,dp = -40\int h\,dh \qquad (7.43)$$

which in turn leads to

$$p^2 + 40h^2 = \text{constant} \qquad (7.44)$$

This is the equation of an ellipse, as shown in Fig. 7.6, and the value of the constant is determined by the values of $p$ and $h$ when the system is initially disturbed. It is clear from this diagram, and from

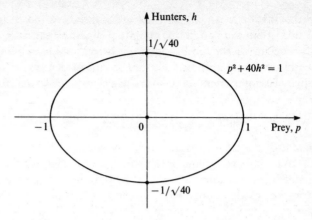

*Figure 7.6.* The elliptical locus of the predator–prey disturbance.

equation (7.44), that both $p$ and $h$ are restricted in value; $p$ and $h$ vary cyclically and the solution $P = 1000$ and $H = 50$ is stable in the sense that slight disturbances do not lead to violent, unrestricted variations – even though the variations are not damped out.

In similar fashion, we can look at the other equilibrium solution – extinction where both the prey and hunter populations $P$ and $H$ are zero (equation (7.40)). Now, with the assumed values of $r = 100$ and $s = 1000$, we find that small disturbances $p$ and $h$ from this extinction solution satisfy the equations

$$\frac{\mathrm{d}p}{\mathrm{d}t} = 100p, \qquad \frac{\mathrm{d}h}{\mathrm{d}t} = -1000h \qquad (7.45)$$

These yield $p = A\mathrm{e}^{100t}$ and $h = B\mathrm{e}^{-1000t}$, where $A$ and $B$ are the initial values of the $p$ and $h$ disturbances. These results show that the prey grows rapidly from zero while the hunters return to extinction: in other words, that extinction is an unstable equilibrium, with the prey achieving rapid growth in the absence of control by the hunters.

This model (based on equations (7.37)), though appealing to mathematicians, is considered unrealistic by most ecologists and biologists, mainly because of the undamped oscillations predicted by the ellipse shown in Fig. 7.6. In nature many cycles are observed, but their period and amplitude do not depend significantly on the initial conditions. They are determined primarily by system parameters such as growth rates and food supply, and usually tend to a stable state (or a stable limit cycle) irrespective of the initial population levels (see for example Fig. 5.5 in Chapter 5). Even so, the model we have looked at provides useful insight into cycling and population collapse. Too many predators over-harvest their prey, causing a decline in the food supply; this is followed by a fall in the number of predators, leading to re-establishment of the prey and a subsequent rise of hunters – and so the wheel goes round and round.

## Equilibrium states and limit cycles

The predator–prey system is unrealistic because perturbations to it give rise to undamped cycling of the populations, determined totally by the starting conditions. An interesting development of the basic predator–prey model is obtained if we suppose that the growth of $P$, the number of prey, is constrained by resources so that, even in the absence of hunters, there is a limit on the prey population. With such a refinement the predator–prey interactions are modified by the incorporation of an extra term $(-P/L)$ in the prey growth equation:

$$\frac{\mathrm{d}P}{\mathrm{d}t} = rP\left(1 - \frac{P}{L}\right) - aHP, \qquad \frac{\mathrm{d}H}{\mathrm{d}t} = (-s + bP)H \quad (7.46)$$

These equations represent a predator population $H$ sustained by a prey population $P$ which in turn is sustained by an environment with a prey-carrying capacity $L$. They might represent an ecosystem containing species of carnivore, herbivore and vegetation (such as foxes, rabbits and grass). By definition, the equilibrium points of this system are those for which $\mathrm{d}P/\mathrm{d}t$ and $\mathrm{d}H/\mathrm{d}t$ vanish simultaneously, so that both populations are steady. When the simultaneous equations (7.46) for $P$ and $H$ are solved with zero time-derivatives, we obtain

(1) $P = 0$ and $H = 0$.
(2) $P = L$ and $H = 0$.
(3) $P = s/b$ and $H = (1 - Ls/b)r/a$.

Although it is not a trivial task, we can examine the stability of these equilibrium solutions by examining the behaviour of small perturbations $p$ and $h$ about each of them in turn, as before. It can be shown that the total extinction of hunters and prey in state (1) is unstable. State (2) represents a condition in which there are no predators $H$, so the prey has reached the carrying capacity imposed by the environment; this turns out to be stable if and only if $bL < s$. Since we must have positive values for $H$ and $P$, state (3) can be ignored unless $bL > s$; in fact state (3) is stable if and only if this inequality holds.

To sum up, we have shown that for this resource-limited form of the predator–prey model there are three steady states, one (and only one) of which is stable. Just which one it is depends on the system constants; the populations converge on state (2) if $s < bL$, and on state (3) otherwise. The total extinction of both hunters and prey – state (1) – is always unstable.

Models of this sort can be enhanced in various ways, for example by adding a component to represent harvesting by man of the predator, prey or both. Furthermore, there are forms of the growth rates which gives rise to *stable limit cycles*. The significance of a stable limit cycle is that it corresponds to a solution in which the populations cycle, as in Fig. 7.6, but the cycle does not depend on the initial population levels. This is a more acceptable view of cyclic behaviour in ecology than is provided by the simple model described in the last section, because ecologists believe nature to be influenced not so much by initial population levels as by the relationships between species in the system. Such models are often based on modifications of equations (7.46) in which the predator equation is of the form

$$\frac{dH}{dt} = sP\left(1 - b\frac{H}{P}\right) \tag{7.47}$$

This replacement represents the resource-limited growth of predator, the limitation being proportional to the density ($P$) of prey.

### Concluding remarks

Our understanding of ecological system dynamics can be improved by the use of mathematical models based on simple but non-linear differential equations. The mathematical and computational properties of these equations lead quickly to topics of some sophistication, such as bifurcation, chaos and limit cycling. The mathematical features depend on the types of non-linearity included in the growth rates used in such models, and can be effectively followed by using the graphics facility of a microcomputer. Even very simple computer programs can adequately illustrate the dynamical behaviour of these systems.

I hope your imagination has been fired by the richness of behaviour exhibited by the illustrations given here; it is very easy to devise others. But herein lies a danger – the mathematician must avoid falling into the trap of believing that complexity of modelling reflects the complexity of nature. Mathematical models are like metaphors:

if taken too literally they lose their point. Population is only one aspect of an ecological system; others are age structure, genetics, mutation and climatic variations. Even so, as in so many fields of investigation, mathematics provides a powerful means for developing new insights into essentially complicated systems.

## Further reading

D. R. Hofstadter, 'Metamagical themas', in *Scientific American*, Nov. 1981.

R. M. May, *Stability and Complexity in Model Ecosystems* (Princeton University Press, 1973).

J. D. Murray, *Nonlinear Differential Equation Models in Biology* (Clarendon Press, 1977).

W. W. Sawyer, *Mathematician's Delight* (Penguin, 1943).

# 8

## BACK TO THE FUTURE:
## MATHEMATICAL LOGIC
## AND COMPUTERS

### MIKE THORNE

### The moral

Car owners often fall into two classes: those who view cars as a means of getting from A to B, and those who spend most weekends up to their elbows in tappets, grease and oil (if you know what a tappet is, you're already in the second class . . . ). Likewise, people who study mathematics can usually be divided into two fairly well-defined groups: those who neither know nor care how the maths they are learning can help with real-world problems, and those who find it hard to get started on the maths without having some idea of the applications of a particular technique or concept.

I ought not to tell you, I suppose, but in fact *this even applies to some real mathematicians* – many of whom have been getting their sums right for years and years. When amongst other mathematicians, for example, mathematical logicians had to suffer a lot of derisive jokes from their colleagues in this connection, for they were regarded as the purest of the pure – at least until quite recently, when the tables were turned in the logicians' favour.

Less than a quarter of a century ago the study of mathematical logic seemed entirely an end in itself, with no obvious or even non-obvious applications. Indeed, a logician who proved something that had clear-cut practical implications would have been viewed by a colleague working in the 'grease monkey' discipline of aerodynamics as a dude in a pin-stripe suit trying to fix his car at the side of the road without rolling up his sleeves.

But computer scientists have changed all that. More and more results from the field of mathematical logic are being applied to problems in computing in ways that no one would have dreamt of thirty years ago. Thanks to the work done by the logicians without regard for applications, computer scientists have been able to push computing to its present exciting state of development.

So, like all the best stories, this chapter has a moral both for politicians and for us ordinary mortals who vote for politicans (or soon will): in the whole range of scientific endeavour – not just mathematics – it is very difficult to determine which new techniques you are going to need tomorrow and which will turn out to be genuinely useless. We should all remember this, and hedge our bets rather than cutting off avenues of research which might seem crazy and obscure today, just in case they turn out to provide the mainstay of scientific activity most relevant to industry tomorrow.

## Introduction

This chapter is about the application of ideas in mathematical logic to computer science. Most recently, computer scientists have turned to mathematical logic (and other ideas from mathematics research) to help them solve the so-called 'software crisis'. Stated succinctly, the software crisis is that while, in real terms, the cost of buying a commercial computer has *fallen* by a factor of 100 000 in a decade, the cost of writing programs for use in industry and commerce has *risen* substantially – despite research efforts to make programming an easier, and therefore cheaper, task.

The one most pressing task for computer scientists is therefore to find cheaper ways of producing software. One obvious approach is to invent new programming languages which are more powerful than those already existing in the sense that far fewer program lines are needed in the new language to achieve the same result as with existing languages. We shall discuss one such language, PROLOG, in this chapter (PROLOG stands for PROgramming in LOGic).

The production of any piece of commercial software begins with a careful specification of exactly what that software should do. This specification is written from the point of view of the people who will use the software (the users), rather than from a programmer's point of view.

Research has shown that, during the development of a large computer program, the longer it takes to spot design errors, the more expensive it is to correct them. The question is, therefore, given a program, how can we be absolutely certain that it will always produce results as intended in its design specification? Apart from wanting to contain software development costs, this would be particularly important, for example, in a program controlling a nuclear reactor in a power station. Consequently, computer scientists would like to be able to *prove* that a given specification satisfies a user's requirements.

One approach to solving this problem is to use a version of the language of logic itself as a programming language. In that way, the body of the program would constitute its own proof of correct working. PROLOG is one of the earliest attempts at such a programming language. It turns out not to be so easy to verify that PROLOG programs do perform according to specification. However, PROLOG has become sufficiently well established to be regarded as a successful new programming language, and we shall describe it in some detail towards the end of this chapter.

Another of the problems related to the software crisis arises in the design of new computers and new programming languages. Thus, when about to design a new computer we may have to choose between giving it one of two possible machine-level instruction sets. In this case we need to know whether the simpler of the two instruction sets will allow us to run the same programs. The electronics will be cheaper to design and produce for the simpler instruction set, and it may be easier for us simple-minded humans to write programs using it, and these programs may execute faster. But if this is at the expense of having a less powerful programming language, we may have to opt for the more powerful instruction set, whatever the cost.

Another problem: suppose we want to rewrite a program because we have bought a new computer and the programming language on the new machine requires certain changes to our original program. This often happens; indeed, different versions of the same programming language resemble dialects in the English language, in which a sandwich might variously be known as a 'butty' or a 'sarny'. We must be sure that the new program matches the old one's performance from its users' point of view. Of course, we could run the old and new programs side by side and compare the results. But we must then be very careful to choose a completely representative set of test data. For applications where safety is the most important consideration, a proof would certainly be the better bet.

Since constructing proofs is expensive (it requires very high-grade human labour and takes a lot of time) we shall discuss later in this chapter whether it is possible to automate the process of proving that two programs will always perform exactly alike. More precisely, we shall discuss whether there exists an algorithm which would allow us to examine any two programs and decide if they have the same effect. Also, we shall consider one of the techniques mathematical logicians have invented which make it possible for the computer itself to construct proofs.

Our subject-matter proper begins with two applications of mathematical logic to important issues a little way away from the mainstream attack on solving the software crisis. First, we see what contribution logic can make to telling whether a given program will ever finish the task given to it. Of course, any *finite* measure of time is largely irrelevant here. In practice we may want to know whether a particular program is likely to stop after 10, 15 or 20 minutes, after 3 days, or even in our lifetime. But the important question here is whether it will *ever* finish, given an infinite amount of time.

Second, can logic tell us whether there are some things no computer could ever do? Quite clearly, your BBC micro can't make us a cup of tea, and my Cray supercomputer can't grow feathers. But the boundary between the seemingly possible, say a computer dreaming up nice recipes for dinner (there is a program for doing this), and the seemingly impossible, say a program to write plays for TV (one does exist, though the plays it produces are hardly Shakespeare) gets less distinct daily.

It's now time for a summary of all this introductory material. Then we can get down to the real work.

### Computer science needs logic

The innate connection between computers and logic will be familiar to anyone who has written a few lines of a computer program in LOGO, BASIC, Pascal or even COBOL. But, using results obtained by researchers in mathematical logic – the erstwhile 'useless' branch of mathematical research – computer scientists have been able to make substantial headway with several problems in computing.

In this chapter we shall discuss the following problems and issues in varying degrees of detail to illustrate how an area of research which apparently has no applications can suddenly become vital to another scientific discipline:

(1) Will a particular program ever finish the task given to it?
(2) Are there some things no computer could ever do?
(3) When are programs and machines equivalent?
(4) Can logic help design a new programming language?
(5) Could a computer be used to prove things automatically?

After a brief introduction to mathematical logic, we shall begin with problem (1). But please remember that, while the very nature of the subject is rigour and precision, we shall not be seeing very much of either. Instead – to return to our car analogy – we shall solve the problem of a flat tyre by telling you that a jack and wheelbrace are used, that the jack holds up the car while you remove the wheel that has the punctured tyre, and that the wheelbrace is for undoing the bolts which attach the wheel to the car. Small but essential details such as the prudent application of your left foot to the wheelbrace to loosen a stiff nut will be skimmed over.

## Mathematical logic

Mathematical logic is the science of the formal principles of reasoning. It is concerned with questions such as: 'How do we represent our reasoning processes mathematically?' and 'How can we check an argument for truth or falsity?' The issue is not whether a particular hypothesis is true or false. We are concerned only with the mechanisms of reasoning: given the truth of A and B, does it follow logically that C is true? For example, if we accept that

**if** *you like Cliff Richard* **then** *you are a wally*

is a valid sentence and that

*you like Cliff Richard*

is also valid, then it must be the case that

*you are a wally*

Mathematical logic is not concerned with whether the person who said

*you like Cliff Richard*

was telling the truth or not, nor whether all Cliff Richard fans are wallies. (According to my old headmaster, C.R. is the only pop singer worth listening to, and we all know that headmasters are never wallies. Boards of governors make certain of that kind of thing!)

Now, until very recently, those relatively few mathematicians working in the field of mathematical logic were regarded, sometimes even by the rest of the mathematical community, as oddballs. Granted, one or two of them – for instance Georg Cantor and Bertrand Russell – had discovered paradoxes and had been able to prove theorems which shattered the very foundations of mathematics. But it became a popular myth that all mathematical logicians did was cause trouble over these foundational issues: you couldn't really apply the mathematics they generated to the physical sciences in the same way that calculus and complex numbers aided our understanding of electricity, or the theory of matrices developed our ideas of crystalline structure.

Let's consider a couple of the paradoxes which were discovered just after the turn of the century.

*Berry's paradox*   There are only a finite number of syllables in the English language, and hence there are only a finite number of statements in English containing fewer than 40 syllables. It follows that there are only a finite number of positive whole numbers which are denoted by an English expression containing fewer than 40 syllables (two million, three hundred thousand, four hundred and ninety-two is such an expression). *Let k be the least positive whole number which is not denoted by an expression in the English language containing fewer than forty syllables.* This sentence, in italics, is an English expression containing fewer than forty syllables, yet it denotes the integer $k$. Where have we gone wrong, if at all?

*Russell's paradox*   A *set* is any collection of objects: for example, the set of all non-negative whole numbers, or the set of hamburger bars in Glasgow. The *members* of a set are the individual objects that make up that set. Now, sets may themselves be members of sets: for example, the set of all sets of whole numbers has sets as its members. Most sets are not members of themselves: the set of hot dogs, for example, is not a member of itself because the set of hot dogs is not a hot dog. However, there are sets which do belong to themselves, such as the set of all sets.

Now, consider the set $A$ of all those sets $X$, such that $X$ is not a member of $X$, and ask the question: 'Is $A$ a member of $A$?' If so, with $A$ being a member of $A$, it is a set which is not a member of itself! On the other hand, if $A$ is not a member of $A$, then $A$ ought to belong to the set of all those sets $X$ such that $X$ is not a member of $X$ – namely, $A$!

We should be quite clear that there are no silly errors in these paradoxes. Rather, they serve to point out deep and fundamental weaknesses in our understanding of the way in which we make deductions mathematically. Trying to plug these and related gaps in our knowledge has occupied mathematical logicians for the past century, and completely satisfactory answers to several important questions have still not been found. And as the work has progressed, 'mainstream' mathematicians working on subjects such as calculus and algebra have been told by the logicians that their results increasingly look as if they are built upon foundations of sand!

Not surprising, then, the quirky reputation of practitioners of mathematical logic – at least until their ideas came to be applied to problems like our five problems in computer science. And now we can begin to see how.

### Problem 1: Will it ever stop?

Deciding whether a program will or will not finish the task it specifies is a difficult problem. To see why, try typing the innocent-looking program fragment

```
while x # 1 do
  if (remainder on dividing x by 2) = 0
  then x := x/2
  else x := 3 * x + 1
```

into your home or school computer, and running it for various values of x. Some of the sequences of values of x generated are given in Table 8.1. Believe it or not, *no one* knows whether this innocent little program terminates for all inputs x > 1.

In fact, logicians have established that in general, according to the currently accepted definition of an algorithm (a mechanical procedure for solving a problem in a finite number of operations), deciding whether or not a program will stop is not possible algorithmically. More precisely, there is no general algorithm which takes as its input a program P, and an initial value x to be given to P, which will print YES if P does stop given input x or NO if it does not.

*Table 8.1.*

| Starting value of x | Sequence of values of x generated |
| --- | --- |
| 1 | 1 |
| 2 | 2, 1 |
| 3 | 3, 10, 5, 16, 8, 4, 2, 1 |
| 4 | 4, 2, 1 |
| 5 | 5, 16, 8, 4, 2, 1 |
| 11 | 11, 34, 17, 52, 26, 13, 40, 20, 10, 5, 16, 8, 4, 2, 1 |

Similarly, there is no algorithm which, given as its input a program P, will decide whether or not P will stop for input zero. Neither is there an algorithm to decide whether, given P, there is any value x so that P stops given x as input; nor one to decide whether, given a program P as input, P stops for every conceivable input value.

Unfortunately, these astonishing theorems are too difficult to consider in any further detail. But they should whet your appetite, both for the power the methods of logic bring to acquiring results about computer science, and for the surprising nature of the results themselves.

### Problem 2: Are there some things no computer could ever do?

This question has several obvious (joke) answers in the affirmative, but we concentrate here on whether there are tasks which you might reasonably expect a computer to be able to do, yet which it actually cannot.

### *Solving equations*

Perhaps one of the earliest ideas we learn about computers is that they can be programmed to solve equations. Certainly, solving

$$6x^2 + 13x + 5 = 0 \tag{8.1}$$

wouldn't give anyone a little familiar with programming in BASIC much trouble (you've probably even worked out the answers in your head).

Now consider using a computer to see whether the equation

$$17x^{91} - 43z^{86} - 6y^2x + 1 = 0 \tag{8.2}$$

can be satisfied by integer values of $x$, $y$ and $z$. This is much harder to solve than the simple quadratic equation (8.1). But suppose we wanted to go even a step further, and ask whether there is a computer algorithm, on which we could base a program, which would work out – given *any* polynomial equation – whether or not it can be satisfied by integer values of the variables in it.

A theorem proved by Davies as recently as 1973 implies that it is not possible to produce an algorithm to determine whether an arbitrary polynomial with integer coefficients has integral roots (assuming that polynomials have arbitrarily many variables and arbitrary degree). If we can't get an algorithm, we certainly can't write a program, so a computer won't help us with this problem – at least in general.

In a similar vein, in 1974 Wang effectively considered solving equations like

$$\sin\left(\sin\left(\sin\left(\sin\left((x + \pi)x - (\pi/91)\right)\right)\right)\right) - x + (72/87)\pi = 0 \quad (8.3)$$

made up by using only $x$, $\pi$ and the rational numbers, and combining them with the operations of addition and multiplication, the sine function and compositions of functions. What we would have liked was an algorithm which took as its input an arbitrary equation made up in this way, and which would tell us for which real values of $x$, if any, the equation could be made true. Wang showed that no such algorithm exists. (This isn't to say that, given a *particular* equation of this type, there won't exist an algorithm to find a value for which it is true. Rather, there isn't one grand algorithm which will work for every such equation.)

### Computer-assisted learning

As a last example of this kind, and perhaps the one most relevant of all to students, consider a teaching program written to offer its user some practice in doing integration. For example the program might type out

```
What is ∫ x/(x² + 1) dx ?
```

For our purposes we don't need to know what this means, nor do we need to be able to work out the answer. It is enough to know that the student using the teaching program would be expected to type in a correct answer. Unfortunately there are several possible correct answers;

$$\tfrac{1}{2}\ln(x^2 + 1) + C \quad \text{and} \quad \tfrac{1}{2}\ln A(x^2 + 1)$$

are just two of them.

Now, the person writing the teaching program has two choices. Either make a list of all possible answers, which is impractical since such a list would have to include an infinite number of answers of the form

$$\tfrac{1}{2}\ln A(x^2 + 1) + C + x - x$$

in which parts of the answer cancel each other out, or store just one or two answer(s) and decide if the given answer comes to the same thing as any of the stored answers.

Unfortunately (or fortunately, depending on which way you look at it), in 1968, in another piece of erstwhile 'useless' mathematical logic, Richardson showed that it is impossible to obtain an algorithm to determine whether, for every value of the unknown $x$, the same value is taken by two arbitrary expressions made up from the rational numbers, the two real numbers $\pi$ and $\ln 2$ together with $x$, using

the operations of addition and multiplication
the sine function
the exponential function
the absolute value function
substitution.

Thanks to Richardson's work we know that neither of the possible approaches to producing our drill and practice program is completely satisfactory.

## Problem 3: Program equivalence

Most readers will have experience of different kinds of programming language. Perhaps the two best-known types are what we shall call flowchart programs and **while** programs. Flowchart programs are representative of the program structure typical of machine language programs (and therefore closely related to the language of binary 0's and 1's a computer's hardware understands). **While** programs are based on the modern idea of structured programming found in the facilities offered by such languages as BBC BASIC, Pascal and ALGOL 68.

Most readers will have seen a flowchart program; an example is shown in Fig. 8.1. We might just as well represent this flowchart program by sets of labelled instructions:

```
10 IF A[J] < = 0 THEN GOTO 50
20 J = J + 1
30 A[J] = A[J] + 4
40 GOTO 10
50 END
```

where, by convention, the computation begins at the instruction labelled 10 and finishes when the END operation is executed.

**Definition** *Flowchart programs* are programs containing only instructions of the types shown in Fig. 8.1.

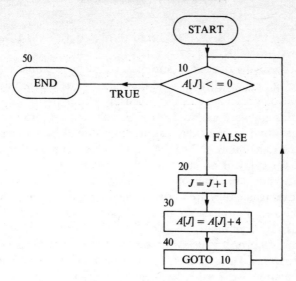

*Figure 8.1.* A flowchart program.

The **while** program equivalent of our flowchart program example is

```
WHILE A[J] > O DO (J := J + 1; A[J] := A[J] + 4)
```

When this is executed, the test for $A[J] > 0$ is made and the instructions in parentheses then get executed if the outcome is TRUE. Then the test is made again and the cycle repeated, execution terminating once the test returns FALSE.

**Definition** *While programs* may contain **while** instructions of the above form, **if** . . . **then** . . . **else** . . . statements, operations such as $J := J + 1$, and the **repeat** ⟨*instructions*⟩ **until** ⟨*condition*⟩ statement, where the ⟨*instructions*⟩ are executed and then the ⟨*condition*⟩ is tested repeatedly until the ⟨*condition*⟩ becomes TRUE. **While** programs do not contain **goto** instructions.

### Why we need to study program equivalence

During the 1970s researchers looked into the sources of error in computer programs. One of the main things they discovered was that errors are more likely to arise when a program is not well-structured: a program based on 'spaghetti logic' would confuse even its own author into making mistakes! Other researchers proposed that much spaghetti logic would be avoided if the **goto** instruction, as found in, for example, BBC BASIC, were thrown out of programming languages altogether.

But if throwing out every **goto** meant that there would be some useful programs which you could no longer write, that would have been a strong argument against such an action. As it turns out, eliminating **goto**'s does result in a slightly weaker programming language – that's one of the reasons why it's still included in BBC BASIC. We shall now outline the proof of this fact as an illustration of what the study of program equivalence can tell us.

In fact, our principal aim is to show that, regardless of the machine on which they are to run, **while** programs can be translated into equivalent flowchart programs. However, the converse is false. The class of computations which can be carried out by **while** programs is a proper subclass of the computations defined by flowchart programs. Thus, to allow for the role of different computers in a computation we need a mathematical model of a computing machine, and it is a very informal description of that to which we now turn.

### *Machines*

A machine supplies all the information missing from a program that is necessary for computations to be described with full mathematical precision. For example, the exact meanings of operations such as $J := J + 1$ and tests such as $A[J] > 0$ will depend on the computer being used. Different computers have different arithmetic capabilities, and $J := J + 1$ might exceed the number range on one computer and not exceed it on another.

An *operation* may be regarded as a transformation of the memory state of a machine, and a *test* corresponds to a truth function which, given a description of the memory state of a machine, will return TRUE or FALSE. There must also exist functions corresponding to the normal input and output operations of a real computer. Thus a formal model of a *machine* M consists of the specification of the following sets and functions:

(1) an input set
(2) a memory set
(3) an output set
(4) an input function $I_M$: input set $\rightarrow$ memory set
(5) an output function $O_M$: memory set $\rightarrow$ output set
(6) for each operation $F$, a function $F_M$: memory set $\rightarrow$ memory set
(7) for each test $T$, a function $T_M$: memory set $\rightarrow$ {TRUE, FALSE}.

Of course, this allows us to define abstract machines which could never exist in real life but which satisfy our mathematical model. Thus we could have a machine with a memory consisting of two calculator displays (or registers), A and B. Input is into register A, and output is from register B. The available operations are $A := A - 1$ and $B := B + 1$, and the only test is '$A = 0$?'. Each of the calculator displays is slightly unusual in that it can store arbitrarily large positive or negative numbers since the memory set is in fact the set of all pairs of integers of the form $(a, b)$.

More formally, we denote the set of positive and negative whole numbers together with zero by $Z$. Then we can define this *two-register machine* as follows:

(1) input set $= Z$
(2) memory set $=$ the set of all vectors with two elements from $Z$, e.g. $(2, -1)$, $(0, 5674)$, $(-1234, -99)$
(3) output set $= Z$
(4) input function $I_M(x) = (x, 0)$ (input into register A)
(5) output function $O_M(x, y) = y$ (output from register B)
(6) operation $A := A - 1$ corresponds to $F_M(x, y) = (x - 1, y)$
(7) operation $B := B + 1$ corresponds to $G_M(x, y) = (x, y + 1)$
(8) test $A = 0$? corresponds to $T_M(x, y) = $ TRUE if $x = 0$, FALSE otherwise.

### Equivalence of programs

To relate programs and machines we need the following definition:

**Definition** P is a program for the machine M if every test and operation appearing in P is specified by M.

Thus the program

```
WHILE A < > 0 DO (A := A - 1; B := B + 1)
```

is a **while** program for the two-register machine which copies the positive number in register A into register B and resets A to zero (try it with A initially containing 6).

Let P and Q be two arbitrary programs, not necessarily of the same type. P is said to be *strongly equivalent* to Q (written P $= =$ Q) if and only if, for the same input, the final outcome from P run on an arbitrary machine M is the same as that obtained from Q when run on M.

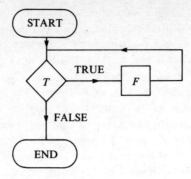

*Figure 8.2.* The program P.

Consider the two programs P (shown in Fig. 8.2) and Q:

```
WHILE T DO (F)
```

P and Q are strongly equivalent, as is evident from their representations.

### The promised theorem

Given any **while** program P we can construct a flowchart program Q
such that P = = Q.

*Informal outline of the proof*    Given a **while** program P, we can build
up a strongly equivalent flowchart Q according to translation rules
of the following type:

(1) Translate an operation *F* as

(2) Translate (**if** *T* **then** *U* **else** *V*) as

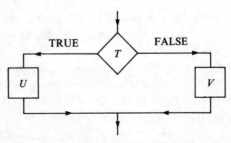

(3) Translate $U; V$ as

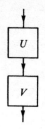

(4) Translate **while** $T$ **do** $(U)$ as

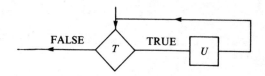

Finally START and END boxes are added in the appropriate places (note that this is not a complete list).

This easy result might give us false hopes about the converse: can every flowchart program be translated into a strongly equivalent **while** program? As we saw earlier, the answer is 'No', as can be demonstrated by providing an example of a flowchart which cannot be thus translated for a careful choice of machine M.

Consider the flowchart F shown in Fig. 8.3, and consider the machine M constructed as follows:

(1) Its input and output sets are both the set of non-negative integers.
(2) Its memory is a single calculator display (register) $X$ capable of holding any non-negative integer, and values are input and output without change.
(3) M has one test, namely $X = 0$?.
(4) M has two operations: $X := X - 1$ and $X := X + 1$.

It follows that running the flowchart program F on this machine will give, as output, 1 if the initial value in $X$ is even, or 0 if the initial value in $X$ is odd. (Check this for yourself by working through the flowchart with some trial values of $X$.)

To begin our proof, we assume that there is some **while** program G, equivalent to F on M. We count the number of distinct instructions in G which involve $X := X - 1$ and suppose there are $k$ of them. Now, consider the computation of G on M with any input $x$ greater than $k$. In order that this initial value may eventually

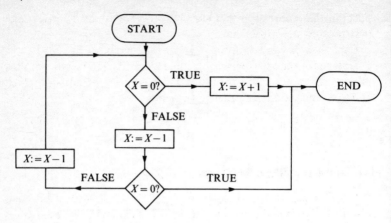

*Figure 8.3.* The program F.

be reduced to 0 or 1, this computation must contain at least $x$ operations of the type $X := X - 1$, each of which is associated with a smaller value of the memory set. It follows that at least one of these operations must occur within the body of some **while** $T$ **do** . . . loop. However, such a loop terminates in the same way for both even and odd $x$, and consequently the computation is unable to distinguish the two cases. Hence G cannot exist.

### Algorithms for deciding program equivalence

There is an algorithm to determine effectively, given two flowchart programs F and G, whether or not F = = G. From this, and the theorem, it follows that, given two **while** programs R and S, we can effectively determine whether or not R = = S. However, to address the problem of program equivalence head on we need the following definition:

**Definition** Two programs F and G are said to be equivalent on a machine M (or *M equivalent*) if, for a given input $X$, whenever F terminates when run on M with input $X$, then running G on M with input $X$ gives exactly the same result – and vice versa.

This definition of the equivalence of programs is the one which is, perhaps, most important practically. It would enable computer designers to discover whether two different possible instruction sets for a new computer led to equivalent programs. However, it turns out that there are very simple machines M for which we can prove that no such algorithm can exist.

But, there *is* an algorithm for deciding, given two flowchart programs for the two-register machine (p. 175), whether or not they are two-register machine equivalent. If operations are allowed which assign a value to a given register which is obtained as a function of the value in another register, then no such corresponding algorithm exists.

### Problem 4: PROLOG – programming in logic

PROLOG was designed in the early 1970s on a theoretical basis established during the previous decade (and earlier) by mathematical logicians. Its creation caused a continuing war in computer science between supporters of languages like BASIC, where the programmer has to say *how* to compute something, and supporters of languages like PROLOG, where the programmer has simply to say *what* to compute. Compare this with the difference between asking a friend who's good at A-level maths to do your homework for you (you only need say *what* to do) and asking a friend who is clever but isn't doing A-level maths (you'll have to go into details of *how* to do it). Of course, if the argument were as simple as that, the *what* brigade would have won easily. (Things weren't that simple, but we don't have space for the details here.)

Among other things, the inventors of PROLOG hoped that it might make it easier to prove that a computer program would always perform according to its specification. Because making PROLOG into a practical reality necessitated departing from pure logic as a programming language, that turned out not to be the case. Also, currently available versions of PROLOG make large demands on computer storage and do not generally execute certain kinds of operation as fast as languages like FORTRAN, so its potential is, as yet, relatively underexploited. Nevertheless, it has found an important place in the set of programming tools available to computer scientists.

A program in PROLOG consists of writing *facts* and *rules*. 'Running' a program involves asking the PROLOG system questions. Thus one might give a PROLOG system the facts

```
likes(john, max_bygraves).
likes(bill, perry_como).
likes(harry, tommy_steele).
```

and the rules

**likes(X, Y) if likes(Y, X).**
**hates(fred, X) if likes(bill, X).**

The facts can be interpreted as

*john likes max_bygraves*
*bill likes perry_como*
*harry likes tommy_steele*

and any rule has two possible types of interpretation. For example, the second rule above could be interpreted either as

*fred hates anybody bill likes*

which is a *declarative* interpretation, or as

*to find someone that fred hates, find someone bill likes*

which is a *procedural* interpretation.

In PROLOG, instead of running a program, we give the system *queries*. Thus the query

**?– hates(fred, perry_como).**

causes the system to print out

**YES**
**?–**

the **?–** signalling that the PROLOG system is once again ready to answer another query. Likewise,

**?– hates(fred, max_bygraves).**

results in the reply

**NO**
**?–**

and

**?– likes(bill, X).**

causes the system to print out all possible values of **X** for which it may be true that **bill likes X**, namely

**X = perry_como**
**No (more) answers**
**?–**

(Notice that whether it is indeed true that **john**, **bill** and **harry** have such dated tastes in music is beside the point. PROLOG simply makes deductions based on the facts it is given. It is the programmer's responsibility to make sure that the 'facts' given to PROLOG are true. In this case I am authorized to tell you that all three people firmly believe that there has never been a better song written than **tommy_steele**'s 'Little White Bull'.)

*Bookworms*

As another example, suppose we now clear the PROLOG system and give it the facts

```
owns(john, hitch_hikers_guide_to_acne).
owns(mary, biggles_gets_a_rough_hairy_chest).
owns(mary, zits_galore).
```

and the rules

```
reads(john, X) if owns(mary, X).
reads(john, X) if owns(john, X).
buys(john, X) if reads(john, X) and not(owns(john, X)).
```

These facts and rules are all examples of what in PROLOG are called *clauses*. In these clauses anything that begins with a capital letter is a variable. We can also have facts in the form

```
owns(john, book(hitch_hikers_guide_to_acne, dougless_spots)).
owns(mary, book(biggles_gets_a_rough_hairy_chest,
                hairy_grower)).
```

which will allow the query

*does john own a book by enid_blyton?*

to be expressed in the form

```
?- owns(john, book(X, enid_blyton)).
```

to which the PROLOG system will respond

```
NO
?-
```

*Which books does john own?* can be expressed either as in the dialogue

```
?- owns(john, X).
X = hitch_hikers_guide_to_acne
No (more) answers
?-
```

or as in the dialogue

```
?- owns(john, book(X, Y)).
X = hitch_hikers_guide_to_acne, Y = dougless_spots
?-
```

*Separating the sexes*

As a final example, we can also have facts like

```
male(john).
male(fred).
female(kim).
```

and the related rules

```
likes(john, X) if female(X).
likes(kim, X) if male(X) and likes(X, cliff_richard).
```

### *Arithmetic in PROLOG*

Arithmetic expressions are often written in a slightly unusual form in PROLOG. Thus, instead of $x + y * z$ we can use

$$+(x, * (y, z)).$$

which more closely resembles PROLOG facts and rules. However, the normal notation is perfectly acceptable. Thus one can have rules such as

```
travelled(fred, Xmiles) if walks(fred, Wdist) and
                    bus(fred, Bdist) and
                    Xmiles is Wdist + Bdist.
```

It is important to realize that arithmetic is relatively unimportant when comparing PROLOG with other programming languages, since PROLOG itself organizes the small execution details which in other programming languages necessitate a considerable amount of arithmetic. Thus, there is no PROLOG equivalent of the FOR . . . NEXT loop structure in BASIC. Repetition in PROLOG is automatic since the system always tries to find every possible solution to a query.

The following arithmetic operations and operators are allowed in PROLOG:
Arithmetic operations

$$X + Y, \quad X - Y, \quad X * Y, \quad X/Y$$

Arithmetic operators

$X = Y$
$X <> Y$   (not equal)
$X < Y$
$X > Y$
$X <= Y$
$X >= Y$

*Rulers of the planet Cruzbon*

We shall illustrate these in relation to clauses for the predicate **reigns**. **reigns(X, Y, Z)** is true if king X reigned on the planet Cruzbon from the year Y to the year Z. (Cruzbon can be seen directly overhead in Wigan, England, at midnight on 28 September in a leap year – it's then actually about 4000 light years from planet Earth. It is easily recognizable because, viewed from that distance, its shape closely resembles that of a corn-flake and its orbit is perceptibly banana-shaped.) Here, then, is our database of kings:

**reigns(bonzo, 1844, 1878).**
**reigns(blanketbrain, 1878, 1916).**
**reigns(pootypops, 1916, 1950).**
**reigns(boris, 1950, 1979).**
**reigns(sushi, 1979, 1991).**

In order to ask who was on the Cruzbon throne in a given year, we ask whether **king(X, Y)** will be true if king X was on the throne in year Y:

**king(X, Y) if reigns(X, A, B) and Y >= A and Y =< B.**

Thus, we can now have the following dialogues with the PROLOG system:

**?– king(sushi, 1991).**
**YES**
**?– king(boris, 1982).**
**NO**
**?–**

*Population densities*

For another example, consider the following facts and rules (to which have been attached some explanatory comments, enclosed

*Table 8.2.* Examples of lists in PROLOG.

| List | Number of elements | head | tail |
|------|--------------------|------|------|
| [a, b, c, d] | 4 | a | [b, c, d] |
| [a] | 1 | a | [ ] |
| [the, cat] | 2 | the | [cat] |
| [ [the, cat], sat] | 2 | [the, cat] | [sat] |
| [the, [dog, ran] ] | 2 | the | [ [dog, ran] ] |
| [the, [cat, sat], down] | 3 | the | [ [cat, sat], down] |
| [ ] | 0 | Undefined | Undefined |

between /* and */):

/* *population figures in millions* */

**population(michigan, 9).**
**population(florida, 8).**
**population(washington, 4).**
**population(hawaii, 1).**

/* *area in 000's of square kilometres* */

**area(michigan, 151).**
**area(florida, 151).**
**area(washington, 176).**
**area(hawaii, 17).**

**density(X, Y) if population(X, P) and area(X, A) and Y is P/A.**

Thus

**?– density(hawaii, 0.059).**
**YES**
**?–**

because the population density of Hawaii is 0.059 million people per thousand square kilometres.

*Lists*

Lists are the means of gathering objects together in PROLOG. The **head** of a list is just its first element: the **tail** of a list **L** is the list you get when you remove the first element of **L**. Some examples are given in Table 8.2.

Lists allow us PROLOG facts of the form

**likes(ted, [john, mary, bill, fred]).**
**telephone_no([john, bloggs], [426,5876]).**

which can be interpreted as

*ted likes john, mary, bill and fred*

and

*john bloggs' telephone number is 426 5876*

We then use variables to get at parts of a list. For example, we may wish to ask if there is anyone whose telephone numbers is 5876:

**?- telephone_no(X, [Y, 5876]).**
**X = [john, bloggs], Y = 426**
**?-**

Given

**suspect(X) if telephone_no(X, [Y, 5876]).**

we obtain the dialogue

**?- suspect([john, bloggs]).**
**YES**
**?-**

Similarly, we can produce a PROLOG program to translate between German and English. Part of the database would be

**means([auf, wiedersehen], [goodbye]).**
**means([die, fernsprechzelle], [telephone, box]).**
**means([vormittags], [in, the, morning]).**
**means([eins], [one]).**

Looking up an entry, as one would in a dictionary, thus becomes

**?- means(X, [in, the, morning]).**
**X = [vormittags]**
**?-**

And finding words associated with another word is just as easy:

**?- means(X, [Y, box]).**
**X = [die, fernsprechzelle], Y = telephone**
**?-**

*Table 8.3.* Accessing members of a list in PROLOG.

| List | Variable settings when matching with X \| Y |
|------|---------------------------------------------|
| **[1, 2, 3, 4]** | **X = 1, Y = [2, 3, 4]** |
| **[ [a, b], [c, d], [e, f] ]** | **X = [a, b], Y = [ [c, d], [e, f] ]** |
| **[a, [b, c], d, 2, 3, 4]** | **X = a, Y = [ [b, c], d, 2, 3, 4]** |

*Table 8.4.* Matching lists in PROLOG.

| List 1 | List 2 | Variables get set to |
|--------|--------|----------------------|
| **[X, Y, Z]** | **[fish, walnuts, cakes]** | **X = fish, Y = walnuts, Z = cakes** |
| **[1, 2, 3, 4]** | **[X, Y \| Z]** | **X = 1, Y = 2, Z = [3, 4]** |
| **[a, b, c, d]** | **[X \| Y, Z, W]** | **X = a, Y = [b], Z = c, W = d** |

## Accessing individual members of a list

Variables can be used to access individual members of a list via a matching process in which the notation X | Y stands for the list which has head X and tail Y, as in Table 8.3. Patterns of this form can be used to match one list with another in various ways, as illustrated in Table 8.4. This table shows, for example, that if PROLOG tries to match the pattern **[X, Y | Z]** against the list **[1, 2, 3, 4]**, then the value of X is set to the same as the first element of the list (X = 1) and then Y | Z is matched against the rest of the list (**[2, 3, 4]**) so that Y = 2, Z = [3, 4].

## List membership

Using the ability to access individual members of a list provided by this pattern matching operation, let us define the **member** predicate which is such that **member(X, Y)** is true if the element X is a member of the list Y:

/* rule 1: *X is certainly a member of any list which has head X* */

**member(X, [X | Y]).**

/* rule 2: *otherwise X is a member of [ Y | Z ] only if X is a member of Z* */

**member(X, [Y | Z]) if member(X, Z).**

Let's see what happens when PROLOG is given the query

   ?– **member(b, [a, b, c]).**

PROLOG tries the rules and facts for **member**, one by one, and in the order given. Thus at first it will try to match

   **member(X, [X | Y]).**

with

   **member(b, [a, b, c]).**

These two will match provided there are no incompatibilities in the settings of the variables: **X = b**, **X = a**, **Y = [b, c]**. But clearly this is impossible: we can't have **X** simultaneously set to **a** and to **b**. Hence PROLOG will now try the next rule, and attempt to match

   **member(X, [Y | Z]).**

with

   **member(b, [a, b, c]).**

which is possible if we take **X = b**, **Y = a**, **Z = [b, c]**. Now, rule 2 for **member** implies that PROLOG should try to prove

   **member(b, [b, c]).**

To do so PROLOG starts once again with rule 1. This time we want to match

   **member(b, [b, c]).**

against

   **member(X, [X | Y]).**

which is possible by setting **X = b**, **Y = [c]**, so that

   **member(b, [a, b, c])**

is indeed true.


### *An argumentative PROLOG program*

As another example, consider writing a program to conduct an argument with the person using the computer. If the user types in

   **[you, are, a, computer]**

we shall arrange for PROLOG to reply

**[i, am, not, a, computer]**

Likewise, if the user types in

**[do, you, speak, german]**

PROLOG will reply

**[no, i, speak, latin]**

To be precise our program will:

(1) accept a sentence typed in as a list
(2) if there are any occurrences of the word 'you' in the sentence, replace each of them with 'i'
(3) arrange that any occurrences of 'are' are replaced by 'am not'
(4) change 'german' to 'latin'
(5) change 'do' to 'no'.

Our program begins with a list of the changes:

**change(you, i).**
**change(are, [am, not]).**
**change(german, latin).**
**change(do, no).**
**change(X, X).**   */* don't change any other words */*

For simplicity we shall expect the user to type in the query in the form

**alter([you, are, a, computer], Y).**

to which the response should be

**Y = [i, am, not, a, computer]**

Here, now, is the definition of **alter** to complete the program:

**alter([ ], [ ]).**
**alter([H | T], [X | Y]) if change(H, X) and alter(T, Y).**

Try to work out what happens step by step during the evaluation of the PROLOG query

**?- alter([excuse, you, german, do, speak], Y).**

*The cut*

When PROLOG is searching for a way to satisfy one of your queries, it will search out every possible solution in a rather unintelligent manner. This may mean that some queries may never be answered, in effect, because the search time would be measured in lifetimes rather than seconds.

The *cut* operator ! can be used by the PROLOG programmer to control the search. Given

**boy(tom).**
**boy(bill).**
**boy(eric).**
**girl(ann).**
**girl(mary).**
**girl(louise).**

the rule

**poss(X, Y) if boy(X) and girl(Y).   /\* *rule 1* \*/**

and the query

**?– poss(X, Y).**

PROLOG would search for all possible matchings of **X** and **Y** to satisfy the query. Thus PROLOG will reply

**X = tom, Y = ann**
**X = tom, Y = mary**
**X = tom, Y = louise**
**X = bill, Y = ann**
   .
   .
   .
**X = eric, Y = mary**
**X = eric, Y = louise**

This method of evaluating the clauses is called *backtracking*. Using the cut interferes with backtracking so that once the cut has been passed, the system is committed to choices already made. Thus, given the rule

**poss2(X, Y) if boy(X) and ! and girl(Y).   /\* *rule 2* \*/**

and the query

**?– poss2(X, Y).**

PROLOG replies

**X = tom, Y = ann**
**X = tom, Y = mary**
**X = tom, Y = louise**

because, once a valid choice of **X** has been made, the PROLOG system cannot choose another.

The cut is one of PROLOG's most controversial features since it effectively destroys any claim that PROLOG is indeed programming in logic. After all, logic is concerned with absolute truths, whereas the effect of a cut is circumstantial. But the fact remains that there are very few useful PROLOG programs which do not contain cuts. Hence, there is scope for new logic programming languages which will execute efficiently on today's computers.

### Problem 5: Using computers to prove things

The introduction to this chapter implied that one of the ultimate aims of computer scientists is to be able to give the computer a list of a user's requirements and a specification for a program, and have the computer prove that the specification meets the user's requirements, no matter what. Research continues towards fulfilling this goal at this very moment. Much of this research derives its inspiration from work mathematical logicians did less than half a century ago which enables us to prove certain kinds of mathematical theorem by means of an algorithmic process.

When implemented on a computer this process works as follows. You type in a collection of assumptions (the *premises*) and the 'fact' (the *hypothesis*) which you think follows from these premises and, provided the hypothesis and premises conform to certain restrictions, the process attempts to construct a *proof by contradiction*. To prove something true by contradiction, we first assume it to be false and try to deduce that something else would then have to be true, something which we know, in fact, to be false. Then, so long as our deduction process was correct, it follows that our assumption must have been incorrect.

For example, suppose we wanted to prove by contradiction that there are infinitely many prime numbers (i.e. whole numbers bigger than 1 which are divisible only by themselves and 1). To do so, we first assume this to be false: that there are only finitely many prime numbers. Let's suppose there are $m$ of them in all:

$$p_1, p_2, p_3, \ldots, p_m$$

Now consider the number

$$X = (p_1 \times p_2 \times p_3 \times \cdots \times p_m) + 1$$

obtained by multiplying them together and adding 1. Since any positive whole number can be written as a product of prime number powers, there must be at least one prime number, call it $p_j$, which is an exact divisor of $X$. But then $p_j$ would be an exact divisor of

$$X - (p_1 \times p_2 \times p_3 \times \cdots \times p_m) = 1$$

which is rubbish because $p_j$ can't simultaneously be an exact divisor of 1 and a whole number bigger than 1! This is the contradiction we have been looking for. Since we've got one, our original assumption – that there were only finitely many prime numbers – must have been false, and so the theorem is proved.

### Resolution

In the above example of proof by contradiction, we repeatedly combined our known facts (the premises) and our 'negated hypothesis' (what we got by assuming the theorem *wasn't* true) until we produced a recognizable contradiction. In order to construct proofs in this way on a computer we need to have an algorithm which tells us how to combine the premises and the negated hypothesis to get a contradiction.

It follows that we need two things:

(1) a technique which allows us to combine any two statements in logic to derive a new logical statement which is in some way simpler; and
(2) an algorithm which tells us which pair of statements to combine next, in the hope that after repeated applications of the combination technique an obvious contradiction is reached.

The technique which allows us to combine two statements in logic to get something simpler is called *resolution*. The idea behind resolution is that if you know that

   *A* **or** *B*

is valid and you know that 'not *A*', written as $\sim A$, is also valid, then you can resolve the two and deduce that *B* must be valid (otherwise *A* **or** *B* would not be valid). Hence

   *A* **or** *B*    and    $\sim A$    resolve to give    *B*

The most easily recognizable contradiction is that both *A* and $\sim A$ are valid. By convention we then say that these can only be resolved to NIL. Hence our algorithm must specify how to combine the premises for a theorem and the negated conclusion using resolution to try to obtain NIL. In fact, the algorithm is a complete cop-out: you just try every single possibility.

We can see how it works on an example. Suppose we have the premises

for any food $X$:   ~(bill eats $X$) **or** john eats $X$
bill eats fish

and the hypothesis

john eats fish

Then, assuming that our hypothesis is not true, all the following must be simultaneously true:

(1)  ~(bill eats $X$) **or** john eats $X$
(2)  bill eats fish
(3)  ~(john eats fish)

We resolve these logic statements in the order given. Resolving (1) and (2) by taking $X =$ fish, we find that the following must be simultaneously true:

john eats fish
~(john eats fish)

Once again, we find that these two resolve, in the order given, to NIL, thus giving us our desired contradiction. Hence, our hypothesis does indeed follow from the premises. (If, however, a contradiction had not been reached at this stage, we would have returned to (1), (2) and (3) and restarted with an attempt to resolve (1) and (3), and so on.)

### *The language of logic – the predicate calculus*

Before the resolution process can begin, the premises and hypothesis must both be expressed in the language of logic – the predicate calculus. Perhaps without realizing it, most people reading this book will have seen statements represented in the predicate calculus. For example,

(all $X$)(($X > 0$) **or** ($X < 0$) **or** ($X = 0$))

which says 'for all values of $X$, $X$ is either 0, less than 0 or greater than 0'. Or

(exists $X$)($X > 42$)

which says 'there exists a value of $X$ which is bigger than 42'. Or

$$((\text{exists } X)(\text{exists } Y)(\text{exists } Z))(X^{42} + Y^{42} + Z^{42} = 0)$$

which says 'there is a value for each of $X$ and $Y$ and $Z$ such that the sum of their 42nd powers is zero'.

In fact, our algorithm for constructing proofs by contradiction using resolution works only when the negated hypothesis and the premises have been put into a standard form, which we call *normal form*. This is necessary because it is possible to state the same logical fact in many different ways. Thus

*sara adores max_bygraves*

and

$\sim \sim$ *(sara adores max_bygraves)*

are logically equivalent statements since the second says that it is not the case that *sara* does not adore *max_bygraves* – meaning that she does!

We insist that all our premises and the negated hypothesis are rewritten in normal form so that they consist of a list of logical statements, all of which must be simultaneously true, and each of which has the form

*A* **or** *B* **or** *C* **or** . . .

thus making it easy to carry out resolution.

To do this we use the equivalences between various logical statements to rewrite the premises and negated hypothesis. For example,

$\sim (A$ **and** $B)$

is equivalent to

$\sim A$ **or** $\sim B$

This is because both these statements are TRUE for exactly the same combinations of truth values of $A$ and $B$, and are both FALSE for exactly the same combinations of truth values of $A$ and $B$. This can be seen from their respective *truth tables*, Tables 8.5 and 8.6. The first two columns in each table range through all the possible combinations of logical values of $A$ and $B$. Since the corresponding entries in the last column of each table are the same, it follows that the two statements are logically equivalent.

By using various logical equivalences, which we shall state at the appropriate time, we shall now describe how to convert a predicate

*Table 8.5.* Truth table for the statement ~(*A* **and** *B*).

| Given this combination of argument values | | The value of *A* **and** *B* is | So the value of ~(*A* **and** *B*) is |
|---|---|---|---|
| *A* | *B* | | |
| TRUE | TRUE | TRUE | FALSE |
| TRUE | FALSE | FALSE | TRUE |
| FALSE | TRUE | FALSE | TRUE |
| FALSE | FALSE | FALSE | TRUE |

*Table 8.6.* Truth table for the statement ~*A* **or** ~*B*.

| Given this combination of argument values | | The value of ~*A* is | The value of ~*B* is | So the value of ~*A* **or** ~*B* is |
|---|---|---|---|---|
| *A* | *B* | | | |
| TRUE | TRUE | FALSE | FALSE | FALSE |
| TRUE | FALSE | FALSE | TRUE | TRUE |
| FALSE | TRUE | TRUE | FALSE | TRUE |
| FALSE | FALSE | TRUE | TRUE | TRUE |

calculus statement into normal form in a series of six steps. Each step will be illustrated by means of a simple example.

*Step 1: Removing implications*

Given

(all *X*)(man(*X*) **implies** has_a_hairy_chest(*X*))

which might be interpreted as 'all men have hairy chests', we rewrite this as

(all *x*)(~ man(*x*) **or** has_a_hairy_chest(*x*))

which does not affect the logical validity of the original statement, since

*A* **implies** *B*

is logically equivalent to

~*A* **or** *B*

*Step 2: Moving negation inwards*

Given the logical equivalence of $\sim (A \text{ and } B)$ and $(\sim A \text{ or } \sim B)$, then

$\sim$ (human(margaret) **and** immortal(margaret))

can be rewritten as

$\sim$ human(margaret) **or** $\sim$ immortal(margaret)

Similarly, we can make use of the equivalence of $\sim (A \text{ or } B)$ and $(\sim A \text{ and } \sim B)$ to move negations so that they are attached to the propositions to which they ultimately apply.

If statements of the form 'all . . .' or 'exists . . .' are present, then a slightly different approach is needed. Thus

$\sim$ (all $Y$)(niceperson($Y$))

('not everyone is a nice person') becomes

(exists $Y$)($\sim$ niceperson($Y$))

using the equivalence of

$\sim$ (all $Y$)($P(Y)$) **and** (exists $Y$)($P(Y)$)

and

$\sim$ (exists $Y$)($P(Y)$) **and** (all $Y$)($\sim P(Y)$)


*Step 3: Removing 'exists'*

We use the following principle: if an object exists with a given property, then give it a name. This technique is known as *Skolemization*, after Skolem, who invented it and demonstrated that, so long as we are careful, then doing this will not lead us to erroneous conclusions about which hypotheses follow from which premises. Thus

(exists $X$)(human($X$) **and** eats($X$, raw_meat))

('there exists a human who eats raw_meat') becomes

human($g1$) **and** eats($g1$, raw_meat)

by giving one of a series of names which may be allocated by this process (in this case let's suppose it's simply $g1$) to the human with the peculiar tastes.

The main problem that can arise here occurs if there are 'all . . .' statements, as in

(all $X$)($\sim$ human($X$) **or** ((exists $Y$)(mother($X$, $Y$))))

('everything is either not human, or there exists someone who is that thing's mother'). If we simply give a name to the mother said here to exist, and then drop the 'exists $Y$ . . . ', we obtain

(all $X$)($\sim$ human($X$) **or** mother($X$, $g2$))

This says that every human has the same mother!

Actually, of course, the mother should depend on the $X$. To make this clear, we use a function $g3(\ )$ to replace the 'exists $Y$ . . .' in this case:

(all $X$)($\sim$ human($X$) **or** mother($X$, $g3(X)$))

('everything is either not human or has a mother, and the name of the mother depends on the thing').

### Step 4: Moving the 'all' expressions outward

Now that our statements do not contain any 'exists . . .' expressions, we can move the 'all . . .' expressions out to the left-hand side of our statements without affecting their logical validity. Thus

(all $X$)($\sim$ man($X$) **or** (all $Y$)($\sim$ dog($Y$) **or** likes($X$, $Y$)))

becomes

(all $X$)(all $Y$)($\sim$ man($X$) **or** ($\sim$ dog($Y$) **or** likes($X$, $Y$)))

Once all the expressions of the form 'all . . .' are on the outside of a statement, then we might as well drop them because we can then assume that any variable has a corresponding 'all . . .'. Thus, the above statement becomes

$\sim$ man($X$) **or** ($\sim$ dog($Y$) **or** likes($X$, $Y$))

It follows that the outcome of the above normalizing steps is to produce logical statements involving just the logical operators **and**, **or** and $\sim$.

### Step 5: Making the **and**'s dominate the **or**'s

We now use the fact that

($A$ **and** $B$) **or** $C$

is equivalent to

($A$ **or** $C$) **and** ($B$ **or** $C$)

and that

  *A* **or** (*B* **and** *C*)

is equivalent to

  (*A* **or** *B*) **and** (*A* **or** *C*)

to get bunches of **or**'s joined together with **and**'s in the middle. Thus

  holiday(*X*) **or** (work(fred, *X*) **and** (angry(fred) **or** sad(fred)))

has the form

  *A* **or** (*B* **and** (*C* **or** *D*))

and is rewritten as

  (holiday(*X*) **or** work(fred, *X*)) **and**
  (holiday(*X*) **or** (angry(fred) **or** sad(fred)))

But we can then drop the brackets amongst the **or**-ed parts to get

  (holiday(*X*) **or** work(fred, *X*)) **and**
  (holiday(*X*) **or** angry(fred) **or** sad(fred))

*Step 6: Putting the expressions into clauses*

We can now abbreviate this further because we can group the **or**-ed parts together as a set of clauses, and we can assume that they are **and**-ed together. Thus, in the above example we get two sets:

  (holiday(*X*), work(fred, *X*))

and

  (holiday(*X*), angry(fred), sad(fred))

and at long last these are indeed what we call 'in normal form'.

Notice that this is exactly what we set out to produce – a collection of logical statements, albeit written in a special way as two sets of clauses, in which each set must be simultaneously true, and the clauses are abbreviations of logical statements of the form

  *A* **or** *B* **or** *C* **or** . . .

*Being clever doesn't mean being able to add up!*

As a complete example of the theorem-proving technique, we'll now show that being clever doesn't necessarily mean that you can add up (or at least it doesn't if you're a monkey!). Let us suppose that

(1) whoever can add up is numerate:
    (all $X$)(addup($X$) **implies** numerate ($X$))
(2) monkeys are not numerate:
    (all $Y$)(monkey($Y$) **implies** $\sim$numerate($Y$))
(3) some monkeys are clever:
    (exists $X$)(monkey($X$) **and** clever ($X$))

and that we wish to prove that

(4) some who are clever cannot add up:
    (exists $Z$)(clever($Z$) **and** $\sim$addup($Z$))

Let us put the premises into normal form, but leave the **or**'s within sets of clauses to make the resolution process clearer:

(1)   addup($X$) **or** numerate($X$)
(2)   monkey($Y$) **or** numerate($Y$)
(3a) monkey($g1$)
(3b) clever($g1$)

Next we must negate the hypothesis (what we want to prove) and put it into normal form:

    $\sim$(exists $Z$)(clever($Z$) **and** $\sim$addup($Z$))
    (all $Z$)($\sim$(clever($Z$) **and** $\sim$addup($Z$)))
    (all $Z$)($\sim$clever($Z$) **or** addup($Z$))

which gives us

(4)  $\sim$clever($Z$) **or** addup($Z$)

Now, we must generate *resolvents*, hoping eventually to produce NIL. If we do, the theory (not shown here) says that our hypothesis does indeed follow from the premises.

    In our example, we get new clauses by resolving as follows:

(5) addup($g1$)     (resolving 3(b) and 4, which we write as [3b, 4])
(6) numerate($g1$)   [5, 1]
(7) monkey($g1$)    [6, 2]

But now we effectively have that $\sim$monkey($g1$) is valid (from 7) and that monkey($g1$) is valid at the same time (from 3(a)). Clearly this has no real logical resolution, so we generate NIL:

(8) NIL       [7, 3a]

Since we've generated NIL, our hypothesis does indeed follow.

The only way in which this worked example differs from a computer implementation of the method is that we have used ingenuity to decide which clauses to resolve so as to get NIL in just a few steps. Since it doesn't have a real brain, let alone any real ingenuity, the computer would have to try all possibilities until it was successful.

### Is there something Marie eats?

As another example, let's solve the following puzzle:

Marie eats whatever Sally eats. Sally eats fish. Is there something Marie eats?

Our premises are thus

(1) (all $X$)(eats(sally, $X$) **implies** eats(marie, $X$))
(2) eats (sally, fish)

and the problem will be solved if we can show that

(3) (exists $X$)(eats(marie, $X$))

Putting 1 and 2 into clausal form gives

(a)  $\sim$ eats(sally, $X$) **or** eats(marie, $X$)
(b)     eats(sally, fish)

Negate the hypothesis:

$\sim$(exists $X$)(eats(marie, $X$))

and put into normal form, giving

(all $X$)($\sim$ eats(marie, $X$))

whence

(c)  $\sim$ eats(marie, $X$)

Thus, resolving, we get

(d)  $\sim$ eats(sally, $X$)     [a, c]
(e)  NIL                 [d, b]   (taking $X = $ fish)

and so we have effectively proved that Marie eats fish.

### The execution of PROLOG programs

Surprise, surprise. PROLOG is an implementation of a resolution theorem prover! When trying to satisfy a PROLOG query, the system takes the query as a hypothesis and tries to prove that it follows from the premises formed by the relevant facts and rules. We negate the query, and resolve the result with the facts, or left-hand sides of the rules, to get new sub-queries which we treat in the same way.

The idea is that

$\sim A$
$A$ if $B$

is in fact the same thing as

$\sim A$
$B$ **implies** $A$

In normal form this is

$\sim A$
$A$ **or** $\sim B$

which we can resolve to get $\sim B$.

To see how this works in practice, consider the PROLOG clauses

**likes( john, bill).**
**likes(tom, X) if likes( john, X).**

These define the predicate **likes**, which, given the query

**?– likes(X, Y).**

will return TRUE if the person **X** likes the person **Y**.

Consider the case

**?– likes(tom, Z).**

which should give $Z =$ **bill**. Negating the query,

$\sim$ **likes(tom, Z).**

and resolving with

**likes(tom, X) if likes( john, X).**

by taking $Z = X$, we get the new sub-query

**?– likes( john, X).**

Again, negate and resolve with

**likes( john, bill).**

by taking $X =$ **bill**. Then we get NIL, so that the initial query is satisfied (since NIL was produced) by taking $Z = X =$ **bill**.

## The way ahead

Great strides have been made with the development of programs which can prove things. And the work continues, because the ultimate goal – a computer mathematician – is an important one. To date, perhaps the most exciting system is Douglas Lenat's AM (Artificial Mathematician) group of programs. AM does much more than just prove theorems: it even proposes theorems to prove. Moreover, it can suggest mathematical concepts which might be worth studying and which might lead to some useful theorems. For example, when given the elementary arithmetic properties of integers, AM suggested that prime numbers would be a fruitful concept to study. Of course, we already know that the concept of a prime number is at the very heart of properties of integers. But AM's discovery shows the terrifying 'intelligence' Lenat's programs embody.

The UK Science and Engineering Research Council (SERC) is running a major new programme of government funded research into the applications of mathematical logic in information technology (IT). (IT covers computer science, but also wider issues such as telecommunications.) According to the SERC, logic represents in software engineering what the calculus represents in mechanical or civil engineering. At last, having served a real-world apprenticeship, mathematical logic has come of age.

What will be next? Well, who knows? When I was an undergraduate, a subject called *category theory* was about the most abstract thing I had to study. It tries to unify various branches of mathematics by seeing them as members of different families. Since the object of study is therefore mathematics itself, it certainly seemed unlikely at the time that category theory would ever have real-world applications. But over the last two or three years, a steady stream of researchers have begun to realize how even that can be of use in solving the software crisis. Which brings us back to our moral. Re-read it, and when you are eligible to vote, act upon it!

## Further reading

R. Bird, *Programs and Machines* (Wiley, 1976).

W. F. Clocksin and C. S. Mellish, *Programming in Prolog* (Springer-Verlag, 1981).

L. Goldschlager and A. Lister, *Computer Science: A Modern Introduction* (Prentice-Hall, 1982).

# 9

## VIBRATIONS: TUNING FORKS AND TRAIN WHEELS, SQUEALING BRAKES AND VIOLINS

### JIM WOODHOUSE

#### Free vibrations and self-sustained vibrations

When you knock on a door or play a note on a piano, you are causing something to vibrate. The vibrating surface makes the surrounding air vibrate, and this carries sound waves to your ears so that you hear a noise. Different types of noise correspond to differences in the nature of the original vibration. We can classify noises according to many different criteria: pleasant or unpleasant, having or not having a musical pitch, carrying on steadily or dying away, and so on. The first of these criteria depends entirely on the particular way our brains process the sound: pleasantness is a subjective notion, not something easily analysed mathematically. The sense of musical pitch is more directly related to the physics of vibrations: it is associated with vibrations which repeat more or less periodically, fast repetition corresponding to high pitch, slower repetition to lower pitch. The natural way to describe a repetition rate, or *frequency*, is in cycles per second, usually now called *hertz* after the nineteenth-century German physicist Heinrich Hertz, and conventionally abbreviated to Hz. As examples, the fastest you can shake your hand is about 5 Hz, middle C on a piano is 262 Hz, and the highest note a human ear can register is about 18 000 Hz, or 18 kHz (kilohertz).

The last of the classification criteria listed above – whether sounds die away or carry on steadily – turns out to correspond to an important distinction in the type of mathematics needed to describe the vibration, which we can explore using some of the ideas introduced in earlier chapters. Both the knock on the door and the piano note are sounds that die away to nothing after a while. The reason for this is not hard to see. In both cases something was set into vibration by a sudden event, then left to its own devices to vibrate in its own natural way. A certain amount of energy was imparted to the vibration by the initial knuckle rap or piano hammer blow, after which no further energy was supplied. Since all moving things dissipate energy to some extent, this initial quota of energy is eventually used up, and vibration ceases. Such vibrations are called *free* or *transient* vibrations. There are many other examples, from the sound of a church bell to the crash of breaking glass.

Some sounds are not like this, though. If you hold down a key on an organ, a note will sound which does not stop until you let go. If you stand under power lines on a windy day, you can hear a 'singing' noise from the wires which changes only when the wind conditions change. If you ride down a hill on your bicycle, and put your brakes on, you may sometimes find that they judder or squeal in an annoying way (judder is a low-frequency vibration, squeal a higher-frequency vibration). If you push a supermarket trolley too fast, you sometimes find that the back wheels start to vibrate vigorously, an effect known as 'wheel shimmy'. All these are examples of *self-sustained* vibration. Since the vibrating object must still be dissipating energy to some extent, there must be a source of energy acting to make up for this and stop the vibration dying away: the organ's blower, the wind, your movement down the hill, or your muscles pushing the trolley. Notice that none of these energy sources is itself vibrating – energy is taken from the steady downhill motion, for example, and somehow produces vibration in the bicycle brakes. To understand how this happens and what governs the vibration frequency and amplitude, we must make a mathematical idealization or *model* of the situation.

### The simplest vibrator

What is the bare minimum of conditions for something to be capable of free vibration? Two things are necessary. There must be a *restoring force*, so that the vibrating object tries to return to its equilibrium position when pushed away from it. And there must be some *inertia*, so that once it gets back to its starting position it does not stop, but overshoots; the restoring force then brings it back again and the cycle repeats, leading to vibration. We would expect the frequency of the vibration to be governed by the balance of restoring force and inertia, and one goal of our mathematical analysis will be to test this expectation.

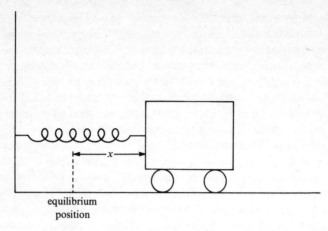

equilibrium
position

*Figure 9.1.* A simple mechanical system – a block on
frictionless wheels, tethered to a wall by a spring.

The simplest mechanical system with these two ingredients is
illustrated in Fig. 9.1. (Incidentally, vibrations need not be mechanical – look at the tuning scale of a radio, and you will see that it is
calibrated in kHz or MHz (megahertz), corresponding to the
frequency of the electromagnetic 'carrier wave' of the radio signal.)
The diagram shows a metal block, of mass 0.2 kg say, which can run
back and forth on a flat surface on frictionless wheels. It is tethered
to a wall by a spring, which can both push (if it is compressed) and
pull (if it is stretched). These elements provide our inertia and our
restoring force, respectively. Let us suppose that the block is displaced away from its equilibrium position by $x$ metres, $t$ seconds
after the moment we choose to start timing events. If we suppose that
the block is set into motion at time $t = 0$ and then left alone to
vibrate, the function $x(t)$ will describe the free vibration.

To find this function we turn to Newton's second law of motion,
which tells us that at any instant the block will be moving in such a
way that the force applied to it by the spring will equal its mass times
its acceleration. Provided the amount of stretching and compression
of the spring remains fairly small, the restoring force will satisfy
Hooke's law (named after the seventeenth-century English scientist
Robert Hooke). This law states that the force of tension is proportional to the displacement, so that it is negative (i.e. compressive,
pushing outwards) when $x$ is negative (to the left) and positive
(tensile) when $x$ is positive (to the right). This is precisely the right
behaviour for a restoring force. The constant of proportionality

measures the strength of the spring: a high value for a stiff spring like
the ones used in a car's suspension system, a lower value for a floppy
spring like the one inside a retractable ball-point pen. As an example,
we take a spring with a constant of proportionality of $1500 \, \mathrm{N \, m^{-1}}$
(which is moderately stiff, about half-way between the car spring and
the pen spring).

We can now write down the mathematical statement of Newton's
second law. We know (from Chapter 1) that the acceleration is the
second derivative of the displacement $x(t)$, so

$$0.2 \, \frac{d^2 x}{dt^2} \; = \; \text{restoring force} \; = \; -1500x \qquad (9.1)$$

The minus sign is necessary because $d^2 x/dt^2$ is the acceleration in
the same direction as $x$, and so is positive to the right, whereas
the restoring force is pulling to the left when $x$ is to the right.
Equation (9.1) is a differential equation which governs the development of the function $x(t)$ as time increases, similar to the ones we met
in Chapter 7, although different in detail.

We can quite easily find possible solutions to this equation in
terms of the familiar sine and cosine functions. If $\omega$ is a constant,
notice that

$$\frac{d}{dt}(\sin \omega t) \; = \; \omega \cos \omega t \qquad (9.2)$$

so that

$$\frac{d^2}{dt^2}(\sin \omega t) \; = \; \omega \frac{d}{dt}(\cos \omega t) \; = \; -\omega^2 \sin \omega t \qquad (9.3)$$

Thus $\sin \omega t$ satisfies equation (9.1) provided $\omega^2 = 1500/0.2 = 7500$,
or $\omega = 86.6$ (in radians per second). By the same reasoning, $\cos \omega t$
also satisfies the equation, with the same value of $\omega$. In fact, because
our differential equation is of second order (i.e. the highest derivative
occurring in it is a second derivative), there are no other independent
(i.e. essentially different) solutions to this equation, a theorem which
is proved in many mathematics textbooks.

The fact that there are no other *independent* solutions does not
mean that there are no other solutions, though. We can make a
whole family of solutions from the two we have by combining them:
you can easily verify that the function

$$x(t) \; = \; A \sin \omega t + B \cos \omega t \qquad (9.4)$$

where $A$ and $B$ are any constants, satisfies equation (9.1). The ability to combine solutions like this is an important property of equations of this type. It is not possible with all differential equations, and the ones for which it is not possible are far harder to solve as a consequence.

If equation (9.4) is the general solution to our differential equation (9.1), it must tell us everything we wanted to know about the free vibration of the block in Fig. 9.1. It certainly describes repetitive motion, since the sine and cosine functions are both *periodic*, with period $2\pi$. Thus $x(t)$ returns to its initial value $x(0)$ when $\omega t = 2\pi$, or $t = 0.0725$ s. This corresponds to vibration at a frequency of $1/0.0725 = 13.8$ Hz. So what does the motion look like? If the block were started by giving it a kick at time $t = 0$, then it will have been given a non-zero speed at that time, but no instantaneous displacement. (Kicking a block does not make it immediately rematerialize in a different place!) It follows that the function we want in this case is $\sin \omega t$, not $\cos \omega t$, since $\sin(0) = 0$ but $\cos(0) = 1$. The motion is then as illustrated in Fig. 9.2(a).

When we first considered how vibration could arise from the combined effects of restoring force and inertia, it seemed plausible that the vibration frequency should be governed by the balance between these two effects. We can now be more specific about that balance. If you rework the calculation above, but with symbolic values $m$ for the mass and $k$ for the spring proportionality constant, you will find that the frequency $f$ (in Hz) of the resulting vibration is given by

$$f = \frac{1}{2\pi}\left(\frac{k}{m}\right)^{1/2} \tag{9.5}$$

The frequency is indeed governed by the ratio of the spring strength (measuring the restoring force) to the mass of the block. Doubling the spring strength would raise the frequency by a factor of $\sqrt{2} = 1.41$, while doubling the mass would lower the frequency by the same factor. In either case, Fig. 9.2(a) would change only in the labelling of the time axis; the sinusoidal form of the vibration would always remain the same.

In the general discussion above, we said that all real objects dissipate some energy when they move, so that free vibrations die away with time. The motion shown in Fig. 9.2(a) does not die away, but that is because we have not included any mechanism for energy loss in our mathematical model. To do so, we would have to include an extra term in our differential equation (9.1) to describe energy dissipation, and then solve the modified equation for the new form

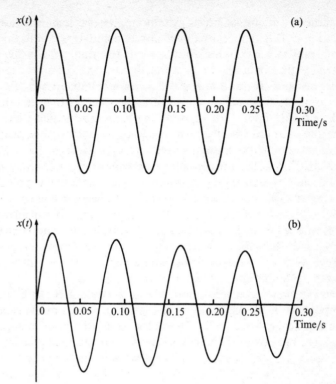

*Figure 9.2.* Oscillatory motion: (a) a constant sine wave;
(b) a decaying sine wave.

of the free vibrations. To go through the details would take us away
from the main task of this chapter, since it would require slightly
more sophisticated mathematics to solve the differential equation,
but we can easily illustrate the result. Figure 9.2(b) shows the effect
of a small amount of energy dissipation on the motion depicted in
Fig. 9.2(a). What we see is essentially the same sinusoidal vibration,
but with an amplitude which decays slowly with time. The form of
this slow decay is precisely the exponential function we met in
Chapter 7, and the rate of decay naturally depends on the magnitude
of the rate of energy loss.

You might well be wondering whether the idealized vibrator
which we have analysed (Fig. 9.1) has any relevance to the real
world. Most things that vibrate do not look at all like blocks rolling
on the ground, restrained by springs. In fact, though, our analysis is
surprisingly general. Although we have considered a particular

idealization of the simplest vibrating system, the differential equation (9.1) describes, to a good approximation, small vibrations of virtually *any* system for which a single function $x(t)$ is sufficient to specify the configuration as a function of time. In other words, diagrams such as those in Fig. 9.2 (with a suitable time scale, of course) apply to small vibrations of a playground swing, a tuning fork, a swing door and many other things. (You might like to consider what provides the restoring force in each of these cases.)

We have not, of course, solved every possible free-vibration problem. The snag is the condition 'for which a single function $x(t)$ is sufficient to describe the configuration'. Consider, for example, the rocking of a car on its suspension when given a downward push on the bonnet. Even if we suppose that the body of the car remains rigid, we need more than one function to describe the motion. A car body can do more than just move vertically up and down in response to such a push, it can also *roll* (about a back–front axis) and *pitch* (about a side-to-side axis). To analyse the behaviour following the push, we would need separate functions to describe these different components of the motion. Perhaps surprisingly, the most efficient mathematical treatment of such problems uses vectors and matrices, introduced in quite a different context in Chapter 2, but that would take us beyond the scope of this brief investigation.

### Why train wheels are conical

Before we move on to discuss self-sustained vibration, we shall look at another, somewhat different example of free vibration. We start from the rather odd-sounding question of why train wheels are the shape they are. Figure 9.3(a) shows a cross-section through a sleeper, the two rails, and one pair of wheels on its axle (a *wheelset*). The wheels are rigidly attached to the axle – they cannot rotate separately. They have *flanges* to stop them coming off the rails, but under normal circumstances one would not want these flanges to rub against the sides of the rails since that would cause rapid wear to both wheels and rails. Such rubbing is discouraged by an ingenious feature of the design of the wheels: notice that the surface of each wheel where it is in contact with the rail is slightly sloping, in opposite directions on the two wheels. Making the wheels slightly conical in this way produces a self-centring action, which we shall now analyse.

It is not hard to get an intuitive feel for this self-centring action. If the wheels were simply cylindrical, with no conical shape, then

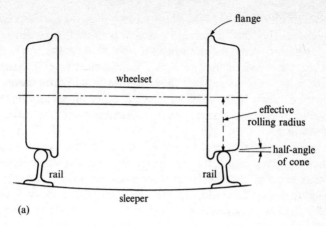

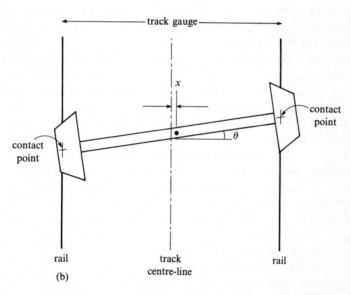

*Figure 9.3.* (a) A wheelset – two train wheels on their axle – resting on rails. (b) Top view, showing the wheelset displaced slightly from its forward direction.

there would be nothing to correct any slight misalignment of the wheelset on the track. It would roll in the direction it was pointing, not quite parallel to the track, until it was stopped by one of the flanges coming into contact with a rail. With conical wheels, on the other hand, something different happens. If the wheelset is not quite centred on the track, one wheel has an effective diameter which is slightly larger than the other because it is running on a different part of the cone. Provided the wheels do not skid on the track, this will make the wheelset tend to turn. With conical wheels as shown in Fig. 9.3(a), this turning is in the right direction to compensate for the misalignment. If the cone were in the other direction, the wheelset would turn to exaggerate the misalignment, which would not be a good idea!

Having seen that conical wheels have a tendency to return towards their central position, this should remind you of restoring forces, and make you ask whether there is then a tendency for vibration of some kind to occur. To investigate this, we shall analyse the simplest case, of a single wheelset rolling along a straight track. (We ignore the rest of the train, with its other wheelsets, bogies, and so on, since to take them into account would be to introduce complications that would make the problem considerably harder to analyse.) A schematic plan view of the wheelset is shown in Fig. 9.3(b). The displacement of the wheelset has been exaggerated for clarity. The centre of the wheelset is displaced a distance $x$ metres from the centre-line of the track, and the axle is rotated by an angle $\theta$ relative to the straight-ahead position. We assume that both $x$ and $\theta$ remain very small throughout the motion, so that we can make approximations to simplify the analysis. For most railways, the *gauge* of the track (the distance between the rails) is 1.435 m. A typical value for the radius of the wheels in the straight-rolling position is 0.42 m. The half-angle of the conical surface of the wheel, shown in Fig. 9.3(a), is commonly 1 in 20, or 1/20th of a radian. We suppose the forward travelling speed of the rolling wheelset to be $20 \, \text{m s}^{-1}$.

Because of the sideways displacement of the wheelset and the cone angle of the wheels, the rolling radii for the two wheels are different. For the left-hand wheel the effective rolling radius is $0.42 - x/20$ metres, while for the right-hand wheel it is $0.42 + x/20$ metres. Consequently, the forward speed at the centre of each wheel differs from the speed $20 \, \text{m s}^{-1}$ at the centre of the axle, by the ratio of these radii to the average radius 0.42 m. Thus the centre of the left-hand wheel is travelling forward at $20[1 - x/(20 \times 0.42)] \, \text{m s}^{-1} = 20(1 - 0.119x) \, \text{m s}^{-1}$. Correspondingly, the centre of the right-hand wheel is travelling forward at $20(1 + 0.119x) \, \text{m s}^{-1}$. The difference between these two speeds causes the wheelset to rotate – in other words it will cause the angle $\theta$ to change with time. The rate of change of $\theta$, $d\theta/dt$, is given (in radians per second) by the difference of the two speeds divided by the distance between the two wheel centres:

$$\frac{d\theta}{dt} = \frac{20(1 + 0.119x) - 20(1 - 0.119x)}{1.435} = 3.32x \quad (9.6)$$

Equation (9.6) gives us one relation between $x$ and $\theta$, but to solve the problem we need another relation between them. Since this one has told us about the rate of change of $\theta$, it seems sensible to ask what governs the rate of change of $x$. The answer to that question is quite simple. The displacement $x$ is changing because the wheelset is running forwards at a speed of $20\,\text{m s}^{-1}$ in a direction which is not quite aligned with the track. The rate of change $dx/dt$, in metres per second, is simply the component of this forward velocity in the direction perpendicular to the track, which is $20\sin\theta$. Thus

$$\frac{dx}{dt} = -20\sin\theta \approx -20\theta \quad (9.7)$$

since $\theta$ is small. (The minus sign is necessary because when $\theta$ is positive, $x$ will be decreasing and vice versa.) We can now combine this with equation (9.6) to obtain an equation of motion for $x$ which does not involve $\theta$ at all. Differentiating both sides of equation (9.7) then substituting from equation (9.6) gives

$$\frac{d^2x}{dt^2} = -20\frac{d\theta}{dt} = -66.4x \quad (9.8)$$

This has precisely the same form as equation (9.1) for the vibrating block, although we have analysed a very different kind of problem this time. Thus we already know the solution: it will be the same as equation (9.4), periodic motion, this time with $\omega = 8.15$, so that the frequency is $8.15/2\pi = 1.30\,\text{Hz}$. The wheelset vibrates from side to side of the track centre-line as it rolls along, in what is called *hunting* motion. In practice, some energy is dissipated via the train's shock absorbers, allowing the hunting vibration to die away, and leading to stable running of the wheelset along the track. Incidentally, if the train is cornering, the same conical wheels and self-centring action allow the curve to be negotiated without either wheel skidding.

### Stick–slide vibration

At the start of this chapter we contrasted free vibration with self-sustained vibration. We can use our simple mechanical system of Fig. 9.1 to illustrate the second type too. Suppose that, instead of making the block move by kicking it, we try to make it vibrate by

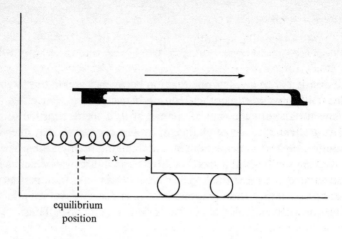

equilibrium
position

*Figure 9.4.* The block from Fig. 9.1, made to vibrate by
bowing it with a violin bow.

pushing a violin bow over it, as shown in Fig. 9.4. We shall assume
a bow speed of $0.5\,\mathrm{m\,s^{-1}}$, a typical value for hand-bowing. This will
produce vibrations, but they will be steady, self-sustained ones which
no longer have the simple sine-wave form of Fig. 9.2(a).

The bow causes vibration through the effect of friction, and we
first need to identify the essential properties of the frictional force
exerted by the bow on the block so that we can incorporate them into
our mathematical model. The force between two bodies in frictional
contact can take either of two different forms, depending on whether
or not one body is sliding relative to the other.

If sliding is occurring, there is a frictional force in the opposite
direction to the sliding (a *drag* force). Its size depends on the force
with which the two bodies are pressed together, and on a physical
constant called the *coefficient of sliding friction* whose value depends
on the nature of the surfaces in contact. The frictional force is then
simply the product of these two factors. For slippery surfaces like a
steel skate on ice the coefficient of sliding friction is small, while it is
much higher for something like rubber sliding on glass. (This is one
reason for putting rubber feet on things which stand on shiny
surfaces: the high coefficient of friction tends to prevent them from
sliding around.) For our example, we suppose that the bow is pressed
against the block with a force of 20 N, and that the coefficient of
sliding friction is 0.3. When sliding is occurring, the frictional force
will thus be $20 \times 0.3 = 6\,\mathrm{N}$, opposing the sliding motion.

When there is no relative sliding, we say that the surfaces are *sticking*. The behaviour of the frictional force is then different. Imagine trying to push a brick across a table top. If you apply only a small force, the brick does not move. The frictional force is obviously balancing the force you are applying. As you push harder, the frictional force increases too, until you reach a critical value when the brick starts to slide. We deduce that the frictional force during sticking can take any value between certain limits. The limit is set by another material constant, the *coefficient of sticking friction*, which is usually greater than the coefficient of sliding friction: in the experiment with the brick on the table, you might notice that once the brick starts to slide, you can keep it moving with a smaller force than you needed to start it. The actual value of the force between these limits is whatever is necessary to prevent sliding from occurring. For our example we shall take the coefficient of sticking friction to be 1.0, so that the frictional force during sticking can take any value between $1.0 \times 20 = 20\,\text{N}$ and $-20\,\text{N}$. For most surfaces, the coefficients of sliding and sticking friction would not be as different as the values 0.3 and 1.0 assumed here. However, these values are quite sensible for a violin bow with rosin on it: we shall soon find out that this large difference is responsible for the ease with which self-sustained vibrations can be created by a violin bow.

The vibration we are aiming to study consists of an alternation between states of sticking and sliding, and is usually called *stick–slide vibration*. To investigate it, we need to consider the sticking and sliding stages separately. Sticking is the simpler one. If the block is sticking to the bow, it is simply moving forward at the same speed as the bow ($0.5\,\text{m s}^{-1}$ here). As it moves, it stretches the spring more and more, so that the restoring force from the spring goes up at a steady rate. This can continue only until that restoring force reaches the limiting value of sticking friction, 20 N. Since the spring force is equal to $1500x\,\text{N}$, where $x$ is the displacement, we must reach limiting friction when $x = 20/1500 = 0.0133\,\text{m}$, or 13.3 mm. After that, the block must begin to slide relative to the bow, and it will presumably spring back to a smaller displacement.

To analyse this sliding stage, we need a differential equation of motion. In fact, we need only make a very simple modification to equation (9.1). The only extra force acting is the frictional force, and we have seen that during sliding this force takes the constant value 6 N. Thus our equation now reads

$$0.2\frac{d^2 x}{dt^2} = -1500x + 6 \qquad (9.9)$$

during sliding. The new term has a positive sign, since the bow is travelling from left to right in Fig. 9.4, so it will pull the block to the right when sliding is occurring. The solution from equation (9.4) needs a correspondingly small modification. Notice that the equilibrium position for motion satisfying equation (9.9) is no longer $x = 0$. If we want it to be possible that $d^2x/dt^2 = 0$, we must move to $x = 6/1500 = 0.004$ m, or 4 mm. This suggests that the mean value of the sine or cosine function satisfying the equation ought to be 0.004 rather than zero, and indeed you can easily verify that the function

$$x(t) = A \sin \omega t + B \cos \omega t + 0.004 \qquad (9.10)$$

satisfies the equation with any constants $A$ and $B$, provided we have the same value $\omega = 86.6$ as on p. 205. Remembering that

$$\sin(\alpha + \beta) = \sin \alpha \cos \beta + \cos \alpha \sin \beta \qquad (9.11)$$

we can rewrite equation (9.10) in the alternative form

$$x(t) = P \sin(\omega t + \phi) + 0.004 \qquad (9.12)$$

where $P^2 = A^2 + B^2$ and $\tan \phi = B/A$. In other words, the general solution to the equation of motion of the block during sliding is sinusoidal motion, with some amplitude $P$ and some *phase angle* $\phi$, about the mean position $x = 0.004$.

We can now put together our knowledge of the separate solutions during sticking and sliding to construct the whole motion, consisting of an alternation between these two states. This is done most easily with the help of a diagram (see Fig. 9.5(a)). Suppose that the block starts from $x = 0$ at time $t = 0$, and that it is initially sticking to the bow. As we have seen, the block then moves forward at the same speed as the bow until it reaches the position $x = 0.0133$, marked by the first black dot on the graph; at that point, sliding begins. We thus have to join the function from equation (9.12) onto the straight line we have already drawn, making suitable choices of the constants $P$ and $\phi$. We have to satisfy two conditions. Since nothing very drastic happens to the block at the moment when sliding begins, both the position $x$ and the speed $dx/dt$ will be the same immediately after the start of sliding as they were immediately before. These two conditions are sufficient to determine the two constants, so there is only one curve that can be drawn, the one shown.

Next we have to decide when sliding ends and sticking starts again. This is simple. If at some later moment the block finds itself moving forward at the same speed as the bow, it will in fact not be

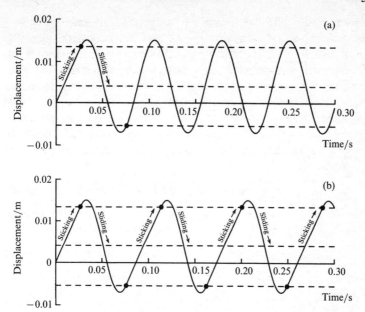

*Figure 9.5.* Motion of the bowed block under (a) a single
'stick–slide' and (b) repeated sticking and sliding.

sliding relative to the bow at all, so it must be sticking again. In other
words, the block follows the trajectory $x(t)$ drawn in Fig. 9.5(a) until
the first point where the slope of the curve (i.e. $dx/dt$) is the same as
it was at the start of sliding. This point is marked by the second black
dot. The bow then 'recaptures' the block, and the sliding curve which
we have plotted beyond that point is irrelevant. Instead, we get what
is shown in Fig. 9.5(b), in which a second episode of sticking is
initiated at the second black dot. The whole cycle can then repeat
indefinitely, sticking and sliding alternately in a completely regular
manner.

In this case, even if we had some energy dissipation in our model,
the motion would still be periodic. The sliding phase of the motion
would consist of part of a cycle of the motion shown in Fig. 9.2(b),
but over such a short time that the decay of vibration amplitude
would not be significant. The loss of energy is made up during the
sticking phase: the bow can do work on the block during sticking.
The longer the period of sticking, the more scope there is for energy
to be fed in to compensate for energy loss during sliding. With the
very different coefficients of sticking and sliding friction assumed so
far, the sticking period is long. But if we choose coefficients which are

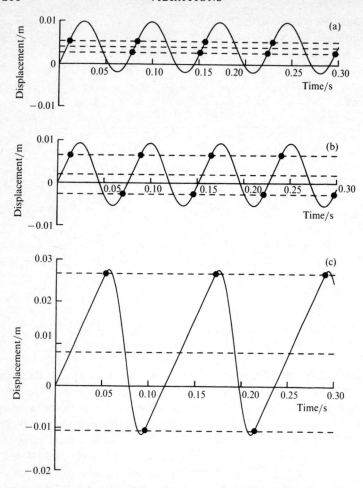

*Figure 9.6.* Stick–slide motion as in Fig. 9.5(b), but with
(a) a lower coefficient of friction between block and bow;
(b) applied bow force half that in Fig. 9.5(b); and (c) applied
bow force double that in Fig. 9.5(b).

closer together, the three dashed lines in the diagram come closer
together, and there is less time for sticking. Figure 9.6(a) shows the
motion under conditions identical to those of Fig. 9.5(b), except that
the coefficient of sticking friction is reduced to 0.4 from its previous
value of 1.0. The curve now looks quite similar to the sinusoidal
shape of Fig. 9.2(a), with only a short interlude of sticking between
episodes of sliding. This gives little opportunity for energy to be fed

into the system, so if there were much energy dissipation during sliding, it would not be possible to sustain this motion. These coefficients of friction are typical of most ordinary surfaces, and this explains why a rosined violin bow is so much better at exciting stick–slide vibration than, say, a metal rod drawn across our moving block.

It is quite interesting to see how the motion of our simple system varies as we change the force with which the bow is pressed against the block. Figures 9.6(b) and (c) show the effects, respectively, of halving and doubling the force relative to that of Fig. 9.5(b), all other conditions remaining the same. The scales on the graphs are the same in each case. Figure 9.6(b), with the lowest force, is somewhat reminiscent of 9.6(a). The motion is almost sinusoidal, with a period just slightly longer than the free period of the oscillator (0.0725 s, derived on p. 206). As we press harder, going via Fig. 9.5(b) to Fig. 9.6(c), sticking occupies an ever-increasing proportion of the period. The amplitude and the period both increase. This tendency can continue indefinitely (at least in our idealized model): by pressing harder and harder, we can make the period as long as we like. Thus the behaviour of this stick–slide oscillator is that with light force it vibrates more or less at the frequency of the free system, and as we press harder the frequency (and thus musical pitch) falls. This is the kind of behaviour that reveals itself in juddering brakes or squeaking door hinges. As you put the brakes on lightly, juddering begins at a frequency governed by the mass and stiffness of the brake mechanism and its mounting. The harder you pull on the brake lever, the slower the juddering becomes.

### Why violin strings are not like juddering brakes

Having considered how our simple vibrator can be excited by a violin bow, it now seems natural for us to ask whether the analysis also applies to a violin string set into vibration by the same bow. A moment's thought will tell you that it is rather a good job that our analysis does *not* apply. When you play a note on a violin, the pitch of the note needs to be rather accurately governed by the length of the string (which you control by stopping with the fingers of your left hand). If the pitch were also sensitive to the force with which the bow is applied, as it was with our vibrating block, then playing a violin in tune would be even harder than it already is.

The reason for the insensitivity of violin notes to bow force is to be found in the very special, indeed rather extraordinary, way in

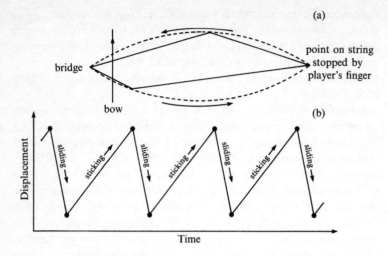

*Figure 9.7.* (a) The actual motion of a violin string:
at any instant of time it shows a sharp 'corner'.
(b) The corresponding 'sawtooth' waveform.

which a bowed string vibrates. This motion was first observed in the
nineteenth century by Hermann von Helmholtz. He was working in
the days before oscilloscopes and other modern laboratory instruments, and his observations were made using ingenious arrangements of rotating mirrors! He found that during a normal, musically
acceptable note the string moves in the way illustrated in Fig. 9.7. At
any instant the string is in two approximately straight portions,
separated by a sharp corner. Two examples are shown in Fig. 9.7(a)
as the two solid lines, representing 'snapshots' at different moments
during the vibration. The sharp corner travels along the dashed line
in the diagram, shuttling back and forth between the two ends of the
string in the directions shown by the two arrows. This dashed line
marks out the 'envelope' of the vibration which is visible to the
naked eye.

Meanwhile, the bow is moving steadily across the string at the
position marked (violins are usually bowed at a point on the string
close to one end, as shown here). What is happening there? As with
the vibrating block in the previous section, we have a stick–slip
vibration. However, the travelling Helmholtz corner plays a vital
role in the timekeeping of the episodes of sticking and sliding, and
this is what makes the behaviour different from that of the block.
For most of the cycle, while the corner travels from the bow to the

right-hand end of the string and back to the bow, the string sticks to the bow. The arrival of the corner dislodges the string, and during the shorter passage of the corner to the left-hand end of the string and back, slipping occurs. Then the corner passes the bow again, triggers recapture and the start of a new sticking period, and the cycle repeats. It follows that at both the instants shown in our two 'snapshots', the string is sticking to the bow. While this 'Helmholtz motion' is going on, the pitch of the note is governed by the travel time of the corner on a round trip along the string and back; it is *not* sensitive to the precise force with which the bow is pressed against the string.

The waveform of displacement at the bowed point on the string is shown in Fig. 9.7(b). It is a *sawtooth wave*, quite reminiscent of Fig. 9.6(c) for the vibration of the block with high bow force. There is the same linear rise during the sticking phase, but this time the 'flyback' during sliding is also a straight line, rather than a segment of sine wave as it was for the block.

This problem is much more difficult to analyse mathematically than the simple vibrator, and so lies rather beyond the scope of this book. The reason is one which we have touched on before. To describe the state of the string at any given instant of time, it is not sufficient to specify just one function. The displacements at different points on the string can be quite different, as the string changes shape during the vibration. The displacement is a function of *more than one variable*: it depends on both time and distance along the string. Even to begin to treat this, we need to extend our ideas of differentiation and differential equations to cope with functions of more than one variable. To do so in sufficient detail to study the *free* motion of the string is not too difficult, but a full mathematical treatment of the self-sustained motion of a bowed string is considerably more complicated, and is indeed still a subject of research work.

We can, however, discuss in a physical way some aspects of the behaviour which are important to violinists. We have just said that the bow force does not influence the pitch of the note played. However, this force does have a significant influence on the sound of the violin in a different way. If you have ever tried playing a violin, you will know that an acceptable musical note can be produced only if the bow force is within certain limits. If you press too hard, you get some kind of raucous 'graunch'. On the other hand, if you do not press hard enough, the note degenerates into a thin, whistling or whining sound which is undesirable. It is often described by players as a 'surface sound'.

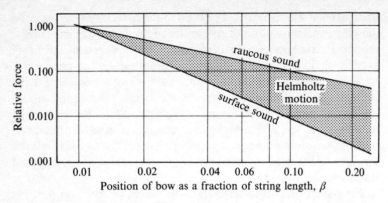

Relative force

Position of bow as a fraction of string length, $\beta$

*Figure 9.8.* Schelleng's diagram: only certain combinations of bowing force and bowing position will produce an agreeable sound from the violin.

But violinists do not control only the bow force. They also have to control two other things during steady bowing: the speed of the bow, and the position of the bow on the string. The influence of bow speed is relatively uninteresting, but the position of the bow on the string is linked in a significant way to the limits on bow force within which a Helmholtz motion can be sustained. This relationship is most easily made clear with the aid of a diagram called *Schelleng's diagram* (Fig. 9.8). To understand this, we need to consider in a little more detail what governs the maximum and minimum bow forces.

If the maximum bow force is exceeded, the string cannot release itself from the bow when the Helmholtz corner arrives. Sticking persists, and there is no transition to sliding. It is not surprising that such a sound would be described as 'raucous'. The period will tend to lengthen, lowering the pitch, and the accurate timekeeping provided by the circulating Helmholtz corner is lost, so that we no longer have a periodic, musical note. Once the force goes above the threshold for this breakdown of the Helmholtz motion, the behaviour of the string may become rather more like that of the simple vibrator, in that further increase of force will progressively decrease the pitch as sticking periods last longer and longer.

If this condition for the maximum bow force is analysed, it turns out that the maximum force depends on the position of the bow on the string. The closer the bow is to one end of the string (the bridge of the violin), the larger the maximum bow force. It is conventional to characterize the bow position by a parameter $\beta$ whose value is the position of the bow as a fraction of the string length, so that $\beta = 0$

corresponds to bowing at the bridge and $\beta = \frac{1}{2}$ corresponds to bowing at the mid-point of the string. In terms of this parameter, it can be shown that the maximum bow force is proportional to $1/\beta$ when $\beta$ is small (as it is for normal playing).

Minimum bow force is governed by something different going wrong with the Helmholtz motion. As we have mentioned several times already, all real vibrating objects dissipate some energy as they vibrate. The violin is certainly no exception, and to sustain the Helmholtz motion the bow must do enough work during the sticking portion of the cycle to compensate for this energy loss. If the bow force gets too low, it can no longer do so, and the string is unable to continue sticking to the bow throughout the long journey of the corner. Somewhere during that time, it slips without being triggered by the passing of the corner. The self-sustained motion then switches over to a different waveform altogether, involving two or more sliding episodes per period. This different waveform produces the characteristic thin, 'surface' sound.

When the condition for this minimum bow force is analysed, it can be shown to depend on the bow position $\beta$ even more strongly than the maximum bow force does. This time, the threshold force turns out to be proportional to $1/\beta^2$. We can now put these two different force limits together in diagrammatic form. It is easiest to plot them on axes calibrated in log (bow force) and log ($\beta$), since both limiting forces then appear as straight lines. The maximum bow-force line has a slope of $-1$, because if force is proportional to $1/\beta$, then log (force) is equal to $-\log(\beta)$ plus a constant. Similarly, the minimum bow-force condition gives a line of slope $-2$. These two lines are shown in Fig. 9.8. The shaded region between them is the region of bow force and bow position for which Helmholtz motion is possible. Outside it are the regions of raucous and surface sounds.

This diagram tells us quite a lot about what is necessary to play a note on a violin successfully. The closer the bow is to the bridge, the larger the force that is necessary, but the narrower the range of acceptable forces. If the bow gets too close to the bridge, the two lines meet, and it becomes impossible to produce Helmholtz motion at all. We can now understand one of the most common errors made by beginners on the violin. Controlling the position of the bow on the string, when you are thinking about many other things such as playing the correct notes on the correct strings, is quite hard. If the bow wanders around on the string, while the bow force remains roughly constant, this corresponds in Fig. 9.8 to moving randomly along a horizontal line. It is obviously quite easy, if this happens, to slip out of the allowed, shaded region. Moving too close to the bridge, you may slip below the minimum bow force; moving too far away may take you above the maximum bow force. Either will lead to some kind of disaster with the sound produced. You might well find that having Schelleng's diagram in mind will make it easier to learn the violin, by explaining to some extent what it is you are supposed to be doing.

### Further reading

R. E. D. Bishop, *Vibration* (Cambridge University Press, 1965).
J. C. Schelleng, 'The physics of the bowed string', in *Scientific American*, Jan. 1974.

# 10

---

## MATHEMATICAL PROGRAMMING: SAVING TIME AND MONEY

### ANDREW NOBLE

### The transportation problem

Bob Brown, the transport manager, sat back in his chair and gazed at the ceiling as if for inspiration. 'Jim,' he said to his assistant, 'this new road-bridge over the Mersey which will be opening soon at Runcorn will affect all our traffic schedules for the delivery of our products. We couldn't use the old transporter bridge because of weight restrictions, so we had to go around by Warrington. Can we achieve a significant reduction in our delivery costs by altering our routes?'

'I don't know,' replied Jim Giles, 'but remember how the people in Planning offered to help us with some new mathematical method they have devised for the computer? At the time we didn't think it was worth the bother, but perhaps it would be different now. I'll give them a ring.'

In industry and commerce, managers often need to investigate a cheaper way of doing something, subject, perhaps, to certain physical limitations or quality standards. Or they may need to organize the production of a manufacturing plant or a whole series of engineering processes so as to yield the most profitable mix of products, bearing in mind customer demand. Often it is not simply the absolute minimum cost or the maximum profit which is of interest, but something approximating to the optimum – what can actually be achieved in the real situation.

It is for these kinds of problem that a particular type of mathematical modelling has been developed. All the techniques used are examples of what is called *mathematical programming*. This has nothing to do with computer programming (although now, of course, computers are employed almost exclusively in making the

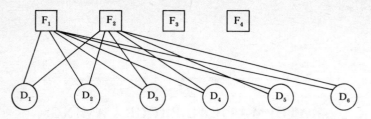

*Figure 10.1.* The transportation problem, for four factories F
and six customers D. (For clarity, routes from factories $F_3$
and $F_4$ are omitted.)

necessary computations). So widely used are these methods that it is
hard to believe that the subject is not yet fifty years old. It was only
in the 1940s that computational rules, or *algorithms*, were devised
which allowed numerical solutions to practical problems to be found
by a sequential process of refinement.

One of the earliest problems to be tackled in this way was that
of supplying many customers from several factories with a single
product. If the cost per unit product supplied along each possible
route from factory to customer is known, which routes should be
used and for how much product?

A simple diagram for four factories and six customers, as shown
in Fig. 10.1, illustrates the problem. It can be seen that there are
$4 \times 6 = 24$ possible routes. In a real problem there are usually
many, many more.

Another way of representing all the information available in a
particular case is to draw up a table like Table 10.1. The first row of
this table means that factory 1 is to deliver 15 units of production to
customer 2, 20 to customer 3, and 10 to customer 4; the second row
shows that factory 2 delivers 40 units to customer 1, and so on. The
last row gives the total requirement of each customer, and the last
column the total production of each factory. (In real life, these two
totals would be unlikely to tally exactly.)

We can make use of this table to consider delivery costs, by
inserting in the top left corner of each cell in the table the cost (in £)
of delivering one unit of production from the factory in question to
the customer in question (see Table 10.2). The total cost is then
found by multiplying together the pair of numbers in each cell, and
adding these products. This gives us £789.

A general form of Table 10.2 is Table 10.3. The mathematical
notation used to describe the elements of a table like this is

*Table 10.1.*

|  |  | \multicolumn{6}{c}{Customers} |  |  |  |  |  |
|---|---|---|---|---|---|---|---|---|
|  |  | $D_1$ | $D_2$ | $D_3$ | $D_4$ | $D_5$ | $D_6$ | Total supply |
| Factories | $F_1$ | 0 | 15 | 20 | 10 | 0 | 0 | 45 |
|  | $F_2$ | 40 | 30 | 0 | 0 | 0 | 0 | 70 |
|  | $F_3$ | 0 | 0 | 0 | 28 | 12 | 10 | 50 |
|  | $F_4$ | 5 | 15 | 10 | 0 | 20 | 25 | 75 |
|  | Total demand | 45 | 60 | 30 | 38 | 32 | 35 | 240 |

*Table 10.2.*

|  |  | \multicolumn{6}{c}{Customers} |  |  |  |  |  |
|---|---|---|---|---|---|---|---|---|
|  |  | $D_1$ | $D_2$ | $D_3$ | $D_4$ | $D_5$ | $D_6$ | Total supply |
| Factories | $F_1$ | 2   0 | 4   15 | 3   20 | 5   10 | 4   0 | 2   0 | 45 |
|  | $F_2$ | 2   40 | 5   30 | 2   0 | 3   0 | 4   0 | 1   0 | 70 |
|  | $F_3$ | 3   0 | 4   0 | 4   0 | 5   28 | 2   12 | 2   10 | 50 |
|  | $F_4$ | 1   5 | 2   15 | 4   10 | 3   0 | 4   20 | 2   25 | 75 |
|  | Total demand | 45 | 60 | 30 | 38 | 32 | 35 | 240 |

potentially rather confusing. When we use a graph or a map we always refer first to the $x$ coordinate (the distance along the horizontal axis) and then to the $y$ coordinate (the distance along the vertical axis). Here the situation is reversed: we refer to the number of the table row first (the 'vertical' axis) and the number of the table column second (the 'horizontal' axis). This is one of those slightly awkward historical conventions. Just remember: row first, column second.

*Table 10.3.*

| | | Customers | | | | | | |
|---|---|---|---|---|---|---|---|---|
| | | $D_1$ | $D_2$ | $D_3$ | $D_4$ | $D_5$ | $D_6$ | Total supply |
| Factories | $F_1$ | $c_{11}$ $X_{11}$ | $c_{12}$ $X_{12}$ | $c_{13}$ $X_{13}$ | $c_{14}$ $X_{14}$ | $c_{15}$ $X_{15}$ | $c_{16}$ $X_{16}$ | $s_1$ |
| | $F_2$ | $c_{21}$ $X_{21}$ | $c_{22}$ $X_{22}$ | $c_{23}$ $X_{23}$ | $c_{24}$ $X_{24}$ | $c_{25}$ $X_{25}$ | $c_{26}$ $X_{26}$ | $s_2$ |
| | $F_3$ | $c_{31}$ $X_{31}$ | $c_{32}$ $X_{32}$ | $c_{33}$ $X_{33}$ | $c_{34}$ $X_{34}$ | $c_{35}$ $X_{35}$ | $c_{36}$ $X_{36}$ | $s_3$ |
| | $F_4$ | $c_{41}$ $X_{41}$ | $c_{42}$ $X_{42}$ | $c_{43}$ $X_{43}$ | $c_{44}$ $X_{44}$ | $c_{45}$ $X_{45}$ | $c_{46}$ $X_{46}$ | $s_4$ |
| | Total demand | $d_1$ | $d_2$ | $d_3$ | $d_4$ | $d_5$ | $d_6$ | $\Sigma s_i$ $= \Sigma d_j$ |

The $c_{ij}$ in the top left of each cell of the table is the cost of sending a unit of product from factory $i$ ($F_i$) to customer $j$ ($D_j$). The capacity of factory $i$ is $s_i$, and the requirement of customer $j$ is $d_j$.

The problem is to satisfy all the customers' requirements for the least total cost. This means finding the quantities of product $X_{ij}$ to be sent by each chosen route such that the sum of all the products $c_{ij}X_{ij}$ is a minimum, i.e. finding the smallest value of

$$c_{11}X_{11} + c_{12}X_{12} + \cdots + c_{45}X_{45} + c_{46}X_{46}$$

The sum of all the $X_{ij}$ in row $i$ equals $s_i$ (i.e. all factory capacity is used). This is a requirement for each row $i$, i.e.

$$X_{i1} + X_{i2} + X_{i3} + X_{i4} + X_{i5} + X_{i6} = s_i, \quad i = 1,2,3,4 \quad (10.1)$$

The sum of all the $X_{ij}$ in column $j$ equals $d_j$ (i.e. all customers are satisfied). This is a requirement for each column $j$, i.e.

$$X_{1j} + X_{2j} + X_{3j} + X_{4j} = d_j, \quad j = 1, \ldots, 6 \quad (10.2)$$

The sum of all the $s_i$ equals the sum of all the $d_j$ (i.e. total supply equals total demand):

$$s_1 + s_2 + s_3 + s_4 = d_1 + d_2 + d_3 + d_4 + d_5 + d_6 \quad (10.3)$$

This is an example of a *transportation linear programming* model ('transportation' because of the type of problem the method was originally devised to solve; 'linear' because all the relationships between the variables $X_{ij}$ are mathematically linear).

But surely this model is simplified to a point where it will be entirely useless in practice? It is clearly quite unrealistic to assume that factory capacity will always exactly equal customer demand – usually supply will exceed demand, or vice versa. However, mathematical modelling is not to be dismissed so easily. The difficulty is overcome by creating imaginary or 'dummy' customers or factories.

If supply exceeds demand a dummy customer is created to whom it costs nothing to send the surplus production from any factory. This dummy customer's requirement is equal to the difference between total supply and total demand. Obviously, in seeking a minimum-cost solution the algorithm will always employ these zero-cost routes in preference to others, so that the entire surplus is always sent to the dummy customer. Similarly, if demand exceeds supply a dummy factory is created with capacity equal to the difference. From this dummy factory it is infinitely expensive to send anything anywhere. (In actual computations 'infinite cost' is substituted by a large positive number.) Such costly routes are never chosen by the algorithm, so the capacity of the dummy factory is never utilized.

The following numerical example illustrates one commonly used method of finding the minimum-cost solution. The basic data ($d_i$, $s_j$ and $c_{ij}$) are set out in Table 10.4 (usually called a *tableau* in mathematical programming jargon). In order to insert the $X_{ij}$, some feasible solution is required which gives the totals $s_i$ and $d_j$, and a rule is necessary to generate such a solution. There are many such rules, one of which is the *north-west corner rule*.

Start in the NW corner of the tableau, and give $X_{11}$ the largest possible value. This will be the lesser of $s_1$ and $d_1$ – in this instance $d_1 = 30$. Column 1 is now complete, and the new NW corner has coordinates (1, 2). The maximum permissible value for $X_{12}$ is $20 = s_1 - 30$. Row 1 is now also complete, and the new NW corner is (2, 2). The maximum value for $X_{22}$ is 30. The new NW corner is now (2, 3). And so on.

The full first feasible solution is shown by the circled numbers in Table 10.4. The cost of this solution is

$$(2 \times 30) + (1 \times 20) + (2 \times 30) + (2 \times 10)$$
$$+ (4 \times 10) + (2 \times 40) + (4 \times 10)$$
$$+ (2 \times 20) + (2 \times 11) = 382 \quad (10.4)$$

*Table 10.4.*

| | $D_1$ | $D_2$ | $D_3$ | $D_4$ | $D_5$ | $D_6$ | Total supply |
|---|---|---|---|---|---|---|---|
| $F_1$ | 2   (30) | 1   (20) | 3   (2) | 3   (4) | 2   (1) | 5   (4) | 50 |
| $F_2$ | 3   (0) | 2   (30) | 2   (10) | 4   (4) | 3   (1) | 4   (2) | 40 |
| $F_3$ | 3   (−2) | 5   (1) | 4   (10) | 2   (40) | 4   (10) | 1   (−3) | 60 |
| $F_4$ | 4   (1) | 2   (0) | 2   (0) | 1   (1) | 2   (20) | 2   (11) | 31 |
| Total demand | 30 | 50 | 20 | 40 | 40 | 11 | 181 |

The question now is, how can this solution be improved to reduce the cost? The computational rule used is called the *stepping-stone algorithm*, for reasons that will soon become clear. The algorithm determines whether it is worth bringing any non-designated route into a new feasible solution. If it is not, the minimum-cost solution has been reached.

For each cell with no $X_{ij}$ in it, there is a 'loop' of potential movement which would bring one unit into that cell. The cost effect round that loop will determine whether such a move should be undertaken.

For each cell $(i, j)$ a loop is defined as follows:

(1) A loop is made up of a number of branches.
(2) A branch connects two cells in the same row or column.
(3) Each branch of the loop is at right angles to the next.
(4) Except for the cell $(i, j)$, the cells on the loop are on the feasible path.
(5) Each cell appears only once.

The loops for cells (4, 1) and (3, 6) are shown in the tableau.

The cost effect round each loop can now be calculated as follows:

(1) Start at cell $(i, j)$, giving its associated cost a positive sign.
(2) Move round the loop, alternating sign between positive and negative.
(3) Compute the algebraic sum of all the costs in the loop.

Thus, for the loops (4, 1) and (3, 6) the results are

$$l_{41} = 4 - 2 + 4 - 4 + 2 - 2 + 1 - 2 = 1$$
$$l_{36} = 1 - 4 + 2 - 2 = -3 \tag{10.5}$$

This means that moving one unit into cell (4, 1) will add to the total cost by 1, but moving one unit into cell (3, 6) will reduce the total cost by 3. All the effects round the relevant loops are also shown in the tableau, in parentheses.

The next step is to choose the cell associated with the largest negative $l_{ij}$ and assign it the maximum possible amount of product. This is equal to the smallest element on the feasible path which is also part of the loop associated with the relevant cell. In the example, cell (3, 6) shows the largest negative effect: $-3$. The smallest element both on the feasible path and on loop (3, 6) is 10, on cell (3, 5).

In order to determine the new reduced-cost feasible solution, and because the tableau must always remain balanced between demand

*Table 10.5.*

|  | $D_1$ | $D_2$ | $D_3$ | $D_4$ | $D_5$ | $D_6$ | Total supply |
|---|---|---|---|---|---|---|---|
| $F_1$ | 2 ㉚ | 1 ⑳ | 3 | 3 | 2 | 5 | 50 |
| $F_2$ | 3 | 2 ㉚ | 2 ⑩ | 4 | 3 | 4 | 40 |
| $F_3$ | 3 | 5 | 4 ⑩ | 2 ㊵ | 4 | 1 ⑩ | 60 |
| $F_4$ | 4 | 2 | 2 | 1 | 2 ㉚ | 2 ① | 31 |
| Total demand | 30 | 50 | 20 | 40 | 30 | 11 | 181 |

and supply, 10 units are

added to cell (3, 6)
deducted from cell (4, 6)
added to cell (4, 5)
deducted from cell (3, 5)

The new feasible solution is shown in Table 10.5. Most of the costs remain unchanged, but for the four cells at the bottom right the total costs are now

$$(2 \times 30) + (1 \times 10) + (2 \times 1) = 72 \qquad (10.6a)$$

compared with

$$(4 \times 10) + (2 \times 20) + (2 \times 11) = 102 \qquad (10.6b)$$

previously. The new solution gives a reduction in cost of 30, as was to be expected from multiplying the effect round the loop (3, 6) by 10.

The whole procedure can now be repeated again and again until no loop effects are negative. For this particular example the eventual solution is shown in Table 10.6. All the cost effects round the loop are now positive, so that no further movement will yield a cost

*Table 10.6.*

| | $D_1$ | $D_2$ | $D_3$ | $D_4$ | $D_5$ | $D_6$ | Total supply |
|---|---|---|---|---|---|---|---|
| $F_1$ | 2 (20) | 1 (30) | 3 (2) | 3 (2) | 2 (0) | 5 (5) | 50 |
| $F_2$ | 3 (0) | 2 (20) | 2 (20) | 4 (2) | 3 (0) | 4 (3) | 40 |
| $F_3$ | 3 (10) | 5 (3) | 4 (2) | 2 (39) | 4 (1) | 1 (11) | 60 |
| $F_4$ | 4 (2) | 2 (1) | 2 (1) | 1 (1) | 2 (30) | 2 (2) | 31 |
| Total demand | 30 | 50 | 20 | 40 | 30 | 11 | 181 |

reduction. The total cost is now

$$(2 \times 20) + (1 \times 30) + (2 \times 20) + (2 \times 20)$$
$$+ (3 \times 10) + (2 \times 39) + (1 \times 11)$$
$$+ (1 \times 1) + (2 \times 30) = 330 \quad (10.7)$$

and this is the minimum cost. The amounts to be transported along each route are ringed in Table 10.6.

Clearly, to carry out these calculations by hand is tedious and prone to error. Nowadays, highly efficient computer packages are available to remove the drudgery.

## Generalized linear programming: a graphic example

The simple transportation problem assumes that each and every factory produces only the same single product. That is not usually the case, especially in some of the more complicated process industries such as oil and chemicals. In these the production processes are highly integrated, any one factory consisting of many linked plants, some of which may produce more than one product. For such factories there is the problem of planning the output of the various products in relation to their market demand so as to make the largest possible 'profit' (the difference between the total costs of production and the total revenue from sales).

Each plant in the factory has a certain capacity. There may be other constraints on the way in which material may flow from plant to plant and from factory to customer. For this type of problem, *generalized linear programming* may be used. This can be illustrated by a simple example.

Imagine a factory which produces two products, X and Y, both of which require high-pressure (HP) steam in their production processes. The capacity for X is 2 units: $x \leqslant 2$; for Y it is 3 units: $y \leqslant 3$. The capacity for HP steam is 12 units; each unit of X requires 4 units of HP steam, each unit of Y requires 3 units, so a further constraint is $4x + 3y \leqslant 12$. The revenue from selling a unit of X is £1000, from a unit of Y it is £2000. The linear programming (LP) problem is, therefore, to maximize profits ($1000x + 2000y$) subject to the various constraints on $x$ and $y$. This we may write as

$$\max (1000x + 2000y)$$

$$\text{subject to} \quad x \qquad \leqslant 2$$
$$y \leqslant 3 \qquad x, y \geqslant 0 \qquad (10.8)$$
$$4x + 3y \leqslant 12$$

The constraints for this simple example may be drawn on a graph, as in Fig. 10.2. They are all straight lines on the graph (hence *linear* programming). The hatching shows on which side of the line lie impermissible values of $x$ and $y$; for example, any value of $x$ to the right of the line $x = 2$ does not satisfy the constraint $x \leqslant 2$. The area bounded by the hatched lines is known as the *feasible region* since all values of $x$ and $y$ in this region satisfy the constraints.

The function to be maximized is $1000x + 2000y$, called the *objective function*. The values of $x$ and $y$ that do this will also give the maximum of $x + 2y$. Imagine, drawn on the graph in Fig. 10.2, a family of parallel lines $x + 2y = k$ for various values of $k$. Two of these are shown: $x + 2y = 2$ and $x + 2y = 6$. These lines cut across the feasible region, and you can see that as the line $x + 2y = k$ moves away from the origin in the direction of the arrow, its final point of contact with the feasible region will be a corner of the feasible region – the point $(\frac{3}{4}, 3)$. The mathematical theory of linear programming shows that the solution always lies at a corner of the feasible region or, in special cases, anywhere along a constraint line joining two corners (when does this happen?). It is

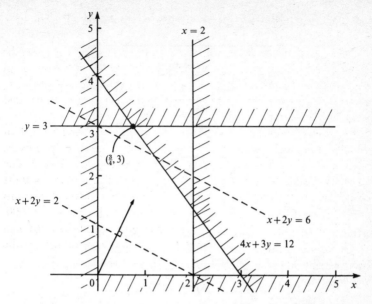

*Figure 10.2.* Example of a generalized linear programming problem. The objective function, which is to be maximized, is $x + 2y$. Two of the parallel lines $x + 2y = k$ are shown. The various constraint lines are shown; the region within them, bounded by the hatching, is called the feasible region.

this that allows the use of a highly efficient computational algorithm called the *simplex method* for problems which involve hundreds of variables and hundreds of constraints in hyperspace (a space of many dimensions).

In the simple problem above the maximum of $x + 2y$ within the feasible region occurs at the corner point $(\frac{3}{4}, 3)$ and has a value $6\frac{3}{4}$. Hence the maximum revenue, £6750, is realized by manufacturing $\frac{3}{4}$ of a unit of X and 3 units of Y. In doing this all the available HP steam is used.

Usually, however, the sensitivity of the result to changes in the variables and constraints is more important than the solution itself. (What would happen if there were more HP steam available? Should the Y plant be extended?) This is because constraints in a mathematical model can never represent exactly the real operating conditions, and because the model itself may leave out of consideration characteristics which are difficult to quantify. While it may seem that the requirement of linearity is a serious limitation to modelling, this is

Table 10.7. The initial simplex tableau for the
problem (10.9).

| Solution variable | Coefficients | | | | | Solution quantity |
|---|---|---|---|---|---|---|
| | $x$ | $y$ | $s_1$ | $s_2$ | $s_3$ | |
| $s_1$ | 1 | | 1 | | | 2 |
| → $s_2$ | | ①  | | 1 | | 3 |
| $s_3$ | 4 | 3 | | | 1 | 12 |
| $C$ | 1 | 2 | 0 | 0 | 0 | 0 |

↑

often not so since, as we shall see later, many non-linear relationships can be linearized or modified in such a way that they can be
analysed by linear programming or other forms of mathematical
programming.

## The simplex method

Quite apart from being a powerful computational algorithm for
arriving at a solution to a generalized linear programming problem,
the simplex method makes it possible to conduct detailed sensitivity
analyses. The method itself will be illustrated with the same simple
example that we solved graphically in Fig. 10.2, namely equations
(10.8), which we can rewrite as

$$\max[(x + 2y) \qquad\qquad\qquad = C]$$
$$x \qquad\quad +s_1 \qquad\qquad = 2$$
$$y \qquad +s_2 \qquad\quad = 3 \qquad (10.9)$$
$$4x + 3y \qquad\qquad +s_3 = 12$$

The variables $s_i$ are called *slack variables*, and can be taken to
represent the 'spare capacity' in the inequalities.

It is now possible to set up the *initial simplex tableau*, arranging
the objective function and the constraints as in Table 10.7. This
tableau shows that

$$s_1 = 2, \qquad s_2 = 3, \qquad s_3 = 12, \qquad C = 0 \qquad (10.10)$$

which of course is a feasible solution $x = 0$, $y = 0$ (at the bottom
left of the feasible region in Fig. 10.2).

*Table 10.8.* The 'intermediate' simplex tableau
for the problem (10.9).

| Solution variable | Coefficients | | | | | Solution quantity |
|---|---|---|---|---|---|---|
| | $x$ | $y$ | $s_1$ | $s_2$ | $s_3$ | |
| $s_1$ | 1 | | 1 | | | 2 |
| $y$ | | 1 | | 1 | | 3 |
| $s_3$ | 4 | 3 | | | 1 | 12 |
| $C$ | 1 | 2 | 0 | 0 | 0 | 0 |

This plan can be improved as follows. Improve the first feasible solution by making as much as possible of the product giving the highest contribution per unit, i.e. the highest figure in the $C$ row. The actual amount of the product will be limited by one or more of the constraints. Hence, in the initial tableau, we select column $y$. Now divide the positive numbers in the $y$ column into the solution quantity column: $2 \div 0$ (which we ignore), $3 \div 1 = 3$ and $12 \div 3 = 4$. Select the row which gives the lowest answer, namely the row identified by $s_2$. Ring the element common to the chosen row and the chosen column: this is the *pivot element*. (Pivot elements arose in a slightly different way in Chapter 2.) Now divide all the elements in the chosen row by the pivot element 1, and change the solution variable to the variable heading the chosen column, i.e. $y$. We now have an 'intermediate' simplex tableau, Table 10.8.

It is now necessary to modify the tableau to reflect the effect of producing three units of Y on all the other rows of the tableau, including the $C$ row. This is done by operating algebraically on each non-pivotal row, including the $C$ row, with a multiple of the pivotal row so that each non-pivotal element in the pivotal column is reduced to zero. Thus three times the second row is subtracted from the third, and two times the second from the fourth, to give the second simplex tableau, Table 10.9. (The $-6$ at the bottom right is in fact a positive contribution of 6; the negative sign arises from the operations of the simplex method.)

The whole process can now be repeated until there are no positive signs in the $C$ row. In this simple case there is only one more iteration required, and the third and final tableau is as shown in Table 10.10. This clearly gives the values $x = \frac{3}{4}$ and $y = 3$, and a total 'profit' or contribution $C = 6\frac{3}{4}$ – exactly the same results as obtained by the graphical method.

*Table 10.9.* The second simplex tableau for the problem (10.9).

| Solution variable | Coefficients | | | | | Solution quantity |
|---|---|---|---|---|---|---|
| | $x$ | $y$ | $s_1$ | $s_2$ | $s_3$ | |
| $s_1$ | 1 | | 1 | | | 2 |
| $y$ | | 1 | | 1 | | 3 |
| → $s_3$ | ④ | 0 | | $-3$ | 1 | 3 |
| C | 1 | 0 | 0 | $-2$ | 0 | $-6$ |
| | ↑ | | | | | |

*Table 10.10.* The third and final simplex tableau for the problem (10.9).

| Solution variable | Coefficients | | | | | Solution quantity |
|---|---|---|---|---|---|---|
| | $x$ | $y$ | $s_1$ | $s_2$ | $s_3$ | |
| $s_1$ | 0 | 0 | 1 | | $-\frac{1}{4}$ | $\frac{5}{4}$ |
| $y$ | 0 | 1 | | 1 | | 3 |
| $x$ | 1 | 0 | | $-\frac{3}{4}$ | $\frac{1}{4}$ | $\frac{3}{4}$ |
| C | 0 | 0 | 0 | $-\frac{5}{4}$ | $-\frac{1}{4}$ | $-6\frac{3}{4}$ |

The evaluation of the slack variables in the objective function can be checked and the total contribution verified by multiplying the slack variable 'costs', $\frac{5}{4}$ and $\frac{1}{4}$, by the amount of the original constraints: $3 \times \frac{5}{4} + 12 \times \frac{1}{4} = 6\frac{3}{4}$. The 'cost' of the other slack variable $s_1$ is zero since there is spare capacity for product X. (Constraints have a 'cost' only when they are fully utilized.)

The coefficients of the slack variables in the objective function are also known as *shadow prices* or *shadow costs*, and they can be used to determine the value of scarce resources. Thus, since $s_2$ has a value of $\frac{5}{4}$, one unit relaxation on the $y$ constraint yields an extra contribution of $\frac{5}{4}$. Similarly, one unit relaxation of the constraint on HP steam yields an extra contribution of $\frac{1}{4}$. This can be verified from Fig. 10.2. However, the relationships hold good only for marginal changes in the constraints.

Simplex tableaux are going to appear again in this chapter, so we are fortunate that there is a straightforward way of simplifying the rather cumbersome presentation given above. The problem is to maximize $z$ where

$$x + s_1 = 2$$
$$y + s_2 = 3$$
$$4x + 3y + s_3 = 12 \quad (10.11)$$
$$z - x - 2y = 0$$

The initial tableau (Table 10.7) may be written with the non-basic variables heading the columns, and basic variables in the rows. The right-hand side of these equations is headed r.h.s. Thus

|       | $x$ | $y$ | r.h.s. |
|-------|-----|-----|--------|
| $s_1$ | 1   |     | 2      |
| $s_2$ |     | ①   | 3      |
| $s_3$ | 4   | 3   | 12     |
| $z$   | $-1$ | $-2$ | 0    |

This is similar to the matrix representations we met in Chapter 2. Subsequent tableaux (Tables 10.9 and 10.10) are

|       | $x$ | $s_2$ | r.h.s. |
|-------|-----|-------|--------|
| $s_1$ | 1   |       | 2      |
| $y$   |     | 1     | 3      |
| $s_3$ | ④   | $-3$  | 3      |
| $z$   | $-1$ | 2    | 6      |

|       | $s_3$ | $s_2$ | r.h.s. |
|-------|-------|-------|--------|
| $s_1$ | $-\frac{1}{4}$ |       | $\frac{5}{4}$ |
| $y$   |       | 1     | 3      |
| $x$   | $\frac{1}{4}$ | $-\frac{3}{4}$ | $\frac{3}{4}$ |
| $z$   | $\frac{1}{4}$ | $-\frac{5}{4}$ | $6\frac{3}{4}$ |

### Extensions of linear and mathematical programming

It is not possible to deal in detail here with the full array of powerful techniques available in mathematical programming; we shall mention only a few. In most linear programming packages available on large computers there are facilities for:

(1) testing the ranges of validity of marginal costs and prices;
(2) determining the range within which a particular cost or price in the objective function does not affect the final solution;
(3) considering changes to a linear programming model outside the range of validity of marginal information. This is called

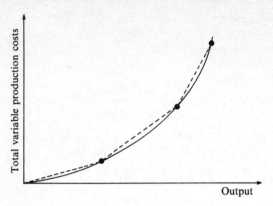

*Figure 10.3*. A convex variable cost curve.

*parametric programming*, and gives the optimal solution for every value of the chosen parameter (variable).

In many production plants unit cost is not constant over the whole feasible range of production. The variable-cost curve frequently takes the shape shown in Fig. 10.3. It arises when the cost of producing a single item increases as more are produced. We say that the curve is *convex*. For our purposes a convex curve is a curve with a positive slope that is increasing smoothly; this means the curve is always below a straight line joining any two points on it.

However, this particular non-linearity can be dealt with easily by linear programming, by dividing the process into three or more processes with constant unit costs represented by the dotted lines. The unit cost of each process is the slope of the relevant line. Since the aim is either to minimize costs or to maximize profits, the process with the lowest unit cost will be employed first, up to its constraint, and then the next, and so on just as though the one continuous convex cost curve were being used. Concave variable-cost curves, which arise when there are economies of scale, cannot be handled by linear programming.

### Non-linearity

So far all the functions considered have been *linear* functions of the variables. It is natural to consider next problems where one or more of these functions (objective or constraints) become non-linear. If the objective function is quadratic and the constraints are linear, an optimum solution can be obtained under certain circumstances by *quadratic programming.* If one or more of the functions is generally non-linear, it may be possible to use *separable programming.* Certain classes of problem require the use of one or more variables which may take only non-negative integer values (including zero). These may be handled by *integer programming* (all variables to be integers) or *mixed integer programming* (only some variables to be integers). Not surprisingly, the difficulty of the mathematical theory, the size of the problem and the time required for computation all increase sharply under these conditions. Any detailed treatment of such methods is outside the scope of this book, but we shall discuss briefly some of the types of problem mentioned above.

### Quadratic programming

Suppose we want to find the maximum and minimum values of the quadratic function $x^2 + y^2$ subject to the condition (or constraint) that $4x + 3y = 12$. Clearly, the solution to this problem (i.e. the values of $x$ and $y$ which give maximum and minimum values for $x^2 + y^2$) consists of points on the straight line $4x + 3y = 12$.

Figure 10.4 shows some of the circles (centred on the origin O) of the family of circles $x^2 + y^2 = r^2$, and also the line $4x + 3y = 12$.

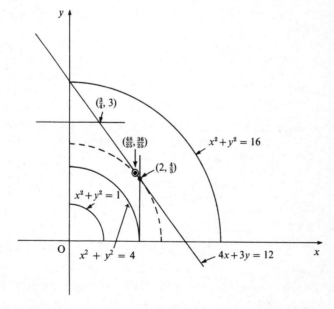

*Figure 10.4.* Example of a quadratic programming problem.

Looking at it, we can see that, in the absence of constraints on the individual values of $x$ and $y$, the minimum value of $x^2 + y^2$ is at the point where $4x + 3y = 12$ is a tangent to one of the family of circles. However, since the absolute values of $x$ and $y$ can increase without limit, then there is no finite maximum. If there were in addition constraints $x \leqslant 2$, $y \leqslant 3$, then of course the maximum value of $x^2 + y^2$ would not be immediately obvious. It must, though, be on the straight line between the points which are its intersections with $x = 2$ and $y = 3$, and the relevant point through which the circle of greatest radius passes is $(\frac{3}{4}, 3)$. This gives a constrained maximum value for $x^2 + y^2$ of 153/16. For the minimum, the line will be a tangent to one of the family of circles where its slope equals the slope of the circle at a point which lies both on the circle and on the line.

The equation of the line may be rewritten as

$$y = -\frac{4}{3}x + 4 \tag{10.12}$$

which shows immediately that the slope is $-4/3$.

The slope of the tangent to the circle $x^2 + y^2 = r^2$ at any point $(x, y)$ on it is found by differentiating the equation with respect to $x$ and solving for $dy/dx$:

$$x^2 + y^2 = r^2 \tag{10.13}$$

$$2x + 2y\frac{dy}{dx} = 0 \tag{10.14}$$

$$\frac{dy}{dx} = -\frac{x}{y} \tag{10.15}$$

For the slope of the tangent and the slope of the line to be equal, $-x/y = -4/3$, i.e. $3x = 4y$. Also, the point must lie on the line $4x + 3y = 12$. Therefore

$$\frac{16}{3}y + \frac{9}{3}y = 12 \tag{10.16}$$

which gives

$$y = \frac{36}{25} \quad \text{and} \quad x = \frac{48}{25} \tag{10.17}$$

Since the point must also lie on the circle $x^2 + y^2 = r^2$, we have

$$\left(\frac{36}{25}\right)^2 + \left(\frac{48}{25}\right)^2 = r^2 \tag{10.18}$$

We can see a connection here with the 3–4–5 right-angled triangle, and that

$$r^2 = \left(\frac{60}{25}\right)^2 = \frac{144}{25} \qquad (10.19)$$

So the minimum value of $x^2 + y^2$ subject to $4x + 3y = 12$ is $144/25$, and it occurs at $x = 48/25$, $y = 36/25$. The solution point (ringed) and the circle (dashed) are shown in Fig. 10.4.

The French mathematician Joseph Louis Lagrange formalized this approach for a wide class of functions and constraints. Further discussion of this method, now called the method of Lagrange multipliers, is beyond the scope of this chapter.

### Integer programming

Many mathematical programming problems require that certain variables take only integer values and, in particular, that variables take only the values 0 and 1. These are known as zero–one variables and may arise in a number of ways.

*Fixed and variable costs*   A factory incurs certain *fixed costs* whatever the level of production. Only if output is zero for a prolonged period can these costs be saved. In addition, production costs vary with output, and are known as *variable costs*. The cost function is in two parts and often looks like Fig. 10.5.

Let the capacity of the factory be $Q$ units of production. Fixed costs $F$ are incurred only if the output ($x$ units) is non-zero. The variable cost of production is $c$ per unit. The total cost is $K$. These conditions can be represented by the following constraints:

$$\left.\begin{array}{c} K = cx + \delta F \\ x \leqslant \delta Q \end{array}\right\} \quad \delta = 0, 1 \qquad (10.20)$$

This idea may be extended to represent generalized step-cost functions.

*Capital costs of new capacity*   Exactly the same situation occurs where there is a choice in an investment programme of whether to build one or more new factories (or make other kinds of investment), or a choice between various sizes for a single new factory.

Capital costs are incurred only if the factory is built. If a single factory is to be built, size to be decided, this condition is expressed by assigning to each possible size a zero–one variable

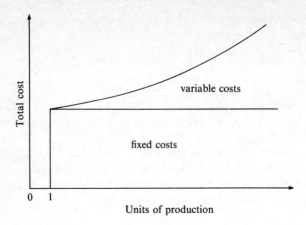

*Figure 10.5.* A cost function made up of fixed and variable costs.

$\delta_i$ $(i = 1, 2, \ldots)$ and requiring that

$$\delta_1 + \delta_2 + \delta_3 + \cdots \leqslant 1 \qquad (10.21)$$

*Manpower scheduling and machine-shop scheduling* Obviously numbers of people must be integral, but duty time is a continuous variable which cannot exceed the total available time. Also, a person is either on duty or off duty, and this may be expressed by a zero–one variable taking the value of one or zero, respectively. In an engineering machine-shop the same is true of machines: they are either busy or idle.

Unfortunately, in the real world, problems which involve integer variables are usually large and complicated and require powerful computers for their solution. Such problems cannot be illustrated here; we must be content with some simple examples to demonstrate the methods. To do so we return to our first problem, equation (10.8), which we can state as

$$\max (z = x + 2y)$$
$$x \qquad \leqslant 2$$
$$y \leqslant 3 \qquad (10.22)$$
$$4x + 3y \leqslant 12$$

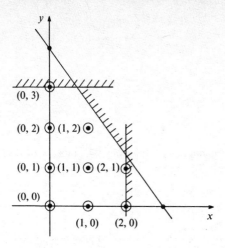

*Figure 10.6.* The problem illustrated in Fig. 10.2 treated as a pure integer programming problem. The feasible region has become a set of discrete points, shown ringed.

but now $x$ and $y$ are required to take integral values only. The feasible region (see Fig. 10.6) is now a set of discrete points. Simple enumeration shows that the maximum point is $(0, 3)$, when $z = 6$. Since the constraints are linear with integral coefficients, the slack variables will also be integral, so this is a *pure integer* programming problem.

The optimal solution in rational numbers may be rearranged, as before, in a tableau:

|  | $s_2$ | $s_3$ | r.h.s. |
|---|---|---|---|
| $z$ | $\frac{5}{4}$ | $\frac{1}{4}$ | $\frac{27}{4}$ |
| $s_1$ |  | $-\frac{1}{4}$ | $\frac{5}{4}$ |
| $y$ | $1$ |  | $3$ |
| $x$ | $-\frac{3}{4}$ | $\frac{1}{4}$ | $\frac{3}{4}$ |

Here $y$ is already an integer (3), but $x = \frac{3}{4}$ is not. From the tableau,

$$x - \tfrac{3}{4}s_2 + \tfrac{1}{4}s_3 = \tfrac{3}{4} \qquad (10.23)$$

This may be rewritten as

$$x + (-1 + \tfrac{1}{4})s_2 + (0 + \tfrac{1}{4})s_3 = 0 + \tfrac{3}{4} \qquad (10.24)$$

Hence

$$\tfrac{1}{4}s_2 + \tfrac{1}{4}s_3 \geqslant \tfrac{3}{4} \tag{10.25}$$

so

$$s_2 + s_3 \geqslant 3 \tag{10.26}$$

from which

$$-s_2 - s_3 + s_4 = -3 \tag{10.27}$$

where $s_4$ is non-negative and integral. This is not satisfied by the current solution, so the new constraint is added to the tableau:

|       | $s_2$          | $s_3$          | r.h.s.          |
| ----- | -------------- | -------------- | --------------- |
| $z$   | $\tfrac{5}{4}$ | $\tfrac{1}{4}$ | $\tfrac{27}{4}$ |
| $s_1$ |                | $-\tfrac{1}{4}$ | $\tfrac{5}{4}$ |
| $y$   | $1$            |                | $3$             |
| $x$   | $-\tfrac{3}{4}$ | $\tfrac{1}{4}$ | $\tfrac{3}{4}$ |
| $s_4$ | $-1$           | $-1$           | $-3$            |

Here $s_3$ enters the tableau in place of $s_4$ since $s_3$ will have the smaller effect on $z$. (Remember we want to move as little as possible away from the original solution with rational numbers.) Then

|       | $s_2$          | $s_4$          | r.h.s. |
| ----- | -------------- | -------------- | ------ |
| $z$   | $1$            | $\tfrac{1}{4}$ | $6$    |
| $s_1$ | $\tfrac{1}{4}$ | $-\tfrac{1}{4}$ | $2$   |
| $y$   | $1$            |                | $3$    |
| $x$   | $-1$           | $\tfrac{1}{4}$ | $0$    |
| $s_3$ | $1$            | $-1$           | $3$    |

Now $x$ is integral as well as $y$, and this is indeed the maximum solution: $x = 0, y = 3, z = 6$.

This method of solution is known as *Gomory's method of integer forms*. First a solution to the problem is found in rational numbers. If these rational numbers are in fact integral no further action is required and the solution is found. If one or more of the variable values is non-integral, a new constraint is added to the current optimal tableau which is satisfied at all feasible non-negative integer values, but not at the current optimal solution. Equation (10.25) above is such a constraint on $x$ and is derived as follows.

Each coefficient in the equation for the relevant variable is split into an integer (negative or non-negative) and a *non-negative* fraction:

$$x + (-1 + \tfrac{1}{4})s_2 + (0 + \tfrac{1}{4})s_3 = 0 + \tfrac{3}{4} \qquad (10.28)$$

$$x + (-1)s_2 + (0)s_3 - 0 = \tfrac{3}{4} - \tfrac{1}{4}s_2 - \tfrac{1}{4}s_3 \qquad (10.29)$$

The left-hand side of this equation is a linear equation in integer variables with integer coefficients, and so is integer-valued at all feasible solutions. Hence the same must be true of the right-hand side. Thus $\tfrac{3}{4}$ and $\tfrac{1}{4}s_2 + \tfrac{1}{4}s_3$ must differ by an integer at all feasible solutions. Now,

$$\tfrac{1}{4}s_2 + \tfrac{1}{4}s_3 \geqslant 0 \qquad (10.30)$$

and

$$\tfrac{1}{4}s_2 + \tfrac{1}{4}s_3 \geqslant \tfrac{3}{4} \qquad (10.31)$$

since if it were less than $\tfrac{3}{4}$, it would follow that

$$\tfrac{1}{4}s_2 + \tfrac{1}{4}s_3 \leqslant \tfrac{3}{4} - 1 = -\tfrac{1}{4} < 0 \qquad (10.32)$$

Therefore

$$\tfrac{1}{4}s_2 + \tfrac{1}{4}s_3 \geqslant \tfrac{3}{4} \qquad (10.33)$$

and this can be written as

$$s_4 - s_2 - s_3 = -3 \qquad (10.34)$$

which is the constraint that was added to the current optimal tableau in rational numbers. Unfortunately the number of steps required to reach an optimum cannot be predicted, and this is clearly a drawback of the method for large problems.

This is an example of what is called a *cutting plane* method, where a cut is made across the feasible region to exclude unacceptable points. The same approach can be adopted for mixed integer problems where some of the variables are rational numbers and the others must be integers. It will not be pursued further here.

## Conclusion

We have looked at the main features of mathematical programming, which is part of the larger mathematical field of optimization. Although mathematical programming has been developed as a powerful technique with many exciting industrial and commercial applications, like any mathematical modelling process it requires skill in problem definition, usually powerful computing, and care and experience in interpreting the results. A distinguished mathematician recently said, 'I have spent all my life doing linear mathematics and that has caused me difficulty enough. Non-linear mathematics is impossible!' Perhaps now you may have just a little insight into what he really meant. Almost all mathematical programming is in fact linear mathematics devised to deal with non-linear situations – amazingly successful for a subject hardly fifty years old.

### Further reading

S. I. Gass, *An Illustrated Guide to Linear Programming* (McGraw-Hill, 1970).

M. S. Makower and E. Williamson, *Operational Research* (Teach Yourself Books, 1975).

D. Smith, *Linear Programming Models in Business* (Polytechnic Publishers, 1973).

K. Williams, *Modern Mathematics for Managers* (Longman, 1972).

# 11

## CRYPTOGRAPHY, OR SENDING
## SECRET MESSAGES

### NELSON STEPHENS

### Sam and Boris and Maxine

Sam is a spy. He has to send messages to his contact, Boris, who works in an embassy. Sam leaves a message in a particular litter-bin in the park at a pre-arranged time each week. So as not to be seen with Sam, Boris picks up the message 15 minutes later.

What can go wrong with their system? Well, Maxine from MI5 removes the message and reads it. She replaces the message so that Sam and Boris do not suspect that their communications are being intercepted and will continue passing messages. That way, Maxine gains new information each week.

There is, of course, a simple remedy available to Sam and Boris. To prevent Maxine from obtaining any information, Sam and Boris agree a secret cipher. Sam enciphers his original message, written in English, and produces an enciphered message which on the face of it looks like a jumble of meaningless characters. It is this enciphered message that he writes down and places in the litter-bin. Boris, in the security of his embassy, deciphers the message according to his secret agreement with Sam, and so reproduces the original message.

But Maxine is no slouch! Not for nothing is she known as the greatest cipher-breaker in MI5. She knows that Sam always begins his messages with 'Dear Boris' and ends with 'With best wishes, Sam.' Armed with such information, she has no problem in breaking the cipher and determining the rest of the message.

This episode belongs more in an old-fashioned spy thriller than in reality. But it highlights accurately a problem in communications: how to retain secrecy when sending a message over an insecure channel. Such a problem arises in many applications. A government communicates with its embassies about very secure matters. A general sends messages to his officers about a plan of campaign. Banks pass financial information to each other; they debit and credit accounts using the ordinary telephone system. The head office of an industrial organization communicates with its branches, sending information which would be desirable to a competitor. A doctor sends a patient's personal data to a hospital. The examples are numerous: in industry, commerce, international politics and the military world, data are passed from one place to another. The means of communication (the *channel*) can be a telephone line, or a letter, or a personal messenger. In none of these instances can total security be guaranteed. The problem facing a *cryptographer* is to devise a way of protecting the information when it is sent over an insecure channel.

Sam is the *encipherer* and Boris is the *decipherer*. The original message is called *plaintext* and the message after it has been enciphered by Sam is called the *ciphertext*. It is the ciphertext that is sent across the insecure channel (the litter-bin in the park) and which the decipherer receives. Maxine is a *cryptanalyst*. Her task is to intercept the ciphertext as it passes over the insecure channel and to use it to determine the plaintext. In our story the encipherer (Sam) and decipherer (Boris) are villains, while the cryptanalyst (Maxine) is the honest girl. In reality, as you will appreciate, the real villain is often the cryptanalyst and the designer of the cipher is the goodie trying to thwart the baddie!

In another scenario, the actors Sam and Boris may be two computers in different cities. Secret information needs to be passed from one computer to the other over a telephone line. This is more secure than a litter-bin, but not without its hazards. Maxine's role is played by some electronic device monitoring the telephone line and sending a copy of the message to a third computer, which is able to analyse the message rapidly and break the cipher. Here the cryptographer has to design a very sophisticated cipher which is secure enough to resist attack by an enemy using every means at his disposal – even perhaps a super-fast computer programmed especially to break ciphers.

I have not yet said very much about the cipher, or in what ways the cryptographer can protect the information, or what tricks are used by the cryptanalyst. You may be wondering what a chapter on cryptography is doing in a mathematics book. Please read on!

We shall start by describing some ciphers used over 2000 years ago, and then move rapidly on to present-day ciphers. The more recent ones have a mathematical background. This is for several reasons: computers are good at arithmetic, performing accurately millions of arithmetic operations in one second, thus allowing plaintext to be enciphered rapidly. More importantly, theoretical mathematics can tell us how difficult it is to solve certain problems quickly. A cipher based on a problem which anyone well-versed in today's mathematics and with access to a modern computer will take a long time to solve has its security underwritten, if not wholly guaranteed. Only an enormous piece of luck on the part of the cryptanalyst, or a major advance in mathematics, or a large increase in the speed of computer arithmetic could render the cipher insecure.

### Some early ciphers

We start by describing a cipher used by Julius Caesar. The English version is to write the letters of the alphabet from A to Z in a line. The encipherer and the decipherer agree a number, $k$, and write the same letters of the alphabet underneath but displaced cyclically by $k$ letters to the left. Thus, when $k = 3$, the tableau looks like

```
ABCDEFGHIJKLMNOPQRSTUVWXYZ
DEFGHIJKLMNOPQRSTUVWXYZABC
```

To encipher a letter in the plaintext, the encipherer finds it in the top line of the tableau and substitutes the letter in the corresponding position in the bottom line:

```
plaintext:    STAN IS IN STAINES
ciphertext:   VWDQ LV LQ VWDLQHV
```

The decipherer, of course, deciphers a letter of the ciphertext by finding it in the bottom line of the same tableau, and substituting the corresponding letter in the top line. A cryptanalyst intercepting the ciphertext may know the kind of cipher being used by the enemy, but not the particular value of $k$ agreed by the encipherer and decipherer for this message.

If our Maxine is responsible for intercepting a modern-day Roman soldier with such a ciphertext, she will have no problem in determining the plaintext. She finds, after trying all 26 possible values for $k$ ($k = 0, 1, \ldots, 25$) that only $k = 3$ gives plaintext that makes any sense. There are two reasons why Maxine is successful. First, she has only a small number of values of $k$ to try. If the number of keys is 26 million, then her task is much larger and she certainly needs to enlist the help of a computer. Second, she is able to reject some of the values of $k$ because the corresponding plaintext does not make sense (remember – Sam's messages are written in English). Maxine is able to exploit the *redundancy* of the language because most collections of letters do not make a word in the language. This second reason works as long as there is enough ciphertext. For example, when the ciphertext is just two consecutive letters, KL, there are two possible corresponding plaintexts, HI when $k = 3$ and NO when $k = 23$, and there is a no way of telling which is the right one. Maxine's task is also impossible if the text has no redundancy to exploit: for example, if the plaintext is just the initials of another spy, almost any combination of letters would be feasible.

At this stage we shall go formal again and state precisely what a cipher is. A *cipher* consists of an *enciphering algorithm* and a *deciphering algorithm*. The encipherer uses the enciphering algorithm, which is just the method stated as a set of instructions, to convert the plaintext to ciphertext. The algorithm uses a *key*, so that the ciphertext depends not only on the plaintext but also on the key.

The decipherer uses the deciphering algorithm to convert the ciphertext into plaintext. It too uses a key. The decipherer's key naturally depends on the encipherer's key, and usually both keys are similar. Knowing one key normally makes it easy to determine the other. Hence, a cryptanalyst will be content in many cases to determine the encipherer's key, for this will yield the decipherer's key and thus the plaintext. With Caesar's cipher, the key is the value of $k$ and the algorithms are just the tableau and the instructions for using it.

Sam and Boris have yet to agree on a cipher. They have rejected Caesar's cipher because it is too easy to break, and they are considering the following cipher. Instead of basing the enciphering algorithm on a fixed rotation of the letters of the alphabet, they consider a tableau where the bottom line of letters is some arbitrary rearrangement of the alphabet. This cipher is called a *permutation* cipher. An example of a tableau might be

```
ABCDEFGHIJKLMNOPQRSTUVWXYZ
GBHKXAVIDZCLRMOQYEWJFUSTPN
```

Once the tableau is agreed between Sam and Boris, enciphering and deciphering is as before. Sam enciphers by substituting the letter on the bottom line for the corresponding letter on the top line. Boris deciphers by substituting the letter on the top line which corresponds to the letter of ciphertext on the bottom line.

For example,

| plaintext: | STAN IS IN STAINES |
| ciphertext: | WJGM DW DM WJGDMXW |

If they think this will foil our super-sleuth Maxine, they are mistaken! Maxine, intercepting the ciphertext, quickly spots that the plaintext contains two English words, each of two letters, and that both begin with the same letter. By a small search of all the combinations, she is left with a small number of possibilities: for example AS AN, or IT IS, or OR ON. These can all be rejected because they do not make the rest of the possible plaintext meaningful – all, that is, except for the two words IS IN (try it). Hence the plaintext is

S**N IS IN S**IN?S

where ** represents two distinct letters and ? represents one letter. With her experience of spies, what they call themselves and where they live, Maxine has no difficulty interpreting the precise plaintext.

Note that Maxine's attack this time is more resourceful. It is impossible for her to try all possible permutations that Sam and Boris might have cooked up because there are

$$26 \times 25 \times 24 \times 23 \times \cdots \times 3 \times 2 \times 1 \; = \; 26!$$

such permutations. Even allowing Maxine the opportunity to try $10^9$ permutations each second (using a computer) it will take her over $10^{10}$ years.

It turns out that there is enough redundancy in the English language to make the job of a cryptanalyst trying to break a permutation code very easy. The systematic approach to breaking such a code is to count the number of occurrences in the ciphertext of each letter, and also of each combination of two and of three letters. Comparing these counts with statistical averages gives enough information for particular letters to be deciphered with a strong probability of success. Once the most common letters are deciphered, the rest can be deciphered either by the same methods (but with less reliability) or by reference to the context and the small number of words which contain the letters already deciphered.

For example, in English text the letter E is expected to occur 13% of the time. The next most common are the letters T, A and O, each occurring about 7% of the time. A letter count of ciphertext immediately shows that the letter occurring most often is the letter enciphered for E. The combination of three letters occurring most often as English plaintext is THE, so this can be used to determine how two further letters of plaintext are enciphered. And so on. All this is effective provided there is enough ciphertext for statistical deductions to be made.

This last point is important. If the plaintext is too short, there is not enough information in the ciphertext for the cryptanalyst to exploit the redundancy in the language. All of this could be made more precise by using a branch of mathematics called information theory. Meanwhile, suffice it to say that Sam and Boris have a problem. If they use a permutation cipher there is a limit to the length of plaintext they can send. They can solve this by changing the permutation (the key) as often as is necessary; but the key consists of 26 letters and is almost as long as the length of the plaintext they can use to send it. In other words, they have to agree a secret key of 26 letters in advance every time they wish to send a message of about the same length. This makes the permutation cipher useless in practice.

So, still searching for a suitable cipher, they are beginning to realize the difficulty of their task. What they require is a cipher which has the following properties:

(1) It must be easy to implement. This means that the algorithms for enciphering and deciphering must be quick and unambiguous.
(2) The key must be significantly shorter than the amount of plaintext which can be enciphered without fear of attack from the cryptanalyst. Preferably, the key should be memorizable. There should, however, be enough possible keys to prevent the cryptanalyst trying to decipher the ciphertext using each of the keys in turn until a sensible message emerges.
(3) It should be secure from other attacks; for example, an attack based on occurrences of certain characters and statistical analysis.

### Block ciphers

This last point means that such a cipher cannot be a simple substitution cipher. Sam and Boris decide that a good cipher must encipher a large group of characters at a time – such a cipher is called a *block cipher*, and works like this. First of all, it is convenient to code the original message into plaintext which has an alphabet of size two: the plaintext is a sequence of zeros and ones. This is achieved by using a conversion code – for example the ASCII code, used to generate characters displayed on computer screens. It has the advantage of reducing every plaintext to a sequence of binary digits (called *bits*), but the disadvantage of increasing the length of the message. Next, the plaintext is blocked. This entails separating it into successive blocks, each block having a fixed number of bits. The cipher acts on one block of bits at a time and produces a block of ciphertext which is the same size as the block of plaintext. Provided the block size is large enough, the statistical attacks described above will be impossible to implement. On the other hand, there must be a simple way of describing how to obtain a block of ciphertext from a block of plaintext, otherwise we are back to a permutation cipher with an even bigger permutation to implement.

When a block of plaintext is enciphered, each bit of the block of ciphertext produced should depend on all the bits of the plaintext block. This property makes an attack more difficult for the cryptanalyst.

To understand how block codes work, Sam and Boris decide they will have to learn some mathematics. It is not very difficult – rather like the arithmetic they learnt at school before they were 11 years old – but necessary if they are to prevent that clever MI5 agent Maxine from reading their secret messages.

### Binary arithmetic

We count using the decimal system. Every positive number is represented by some combination of the ten decimal digits. Thus

$$3257 = 3 \times 10^3 + 2 \times 10^2 + 5 \times 10^1 + 7 \times 1$$

It is well known that computers do their arithmetic using the binary system, where each number is represented by a combination of just two binary digits (i.e. bits). The powers of 2, namely 1, 2, 4, 8, 16, 32, . . ., play the same role as the powers of 10 in decimal arithmetic. A sequence of bits represents a number in the same way. For example, the binary number 1011001 is, in base 10,

$$(1 \times 64) + (0 \times 32) + (1 \times 16) + (1 \times 8)$$
$$+ (0 \times 4) + (0 \times 2) + (1 \times 1) = 89$$

In many cipher systems, a block of bits is used to represent a positive integer. Thus, a block of $b$ bits can represent an integer between 0 and $2^b - 1$ inclusive; conversely, every integer in this range may be represented by a block of $b$ bits.

## Modular arithmetic

Let $m$ be a fixed positive integer. Two integers $x$ and $y$ are then said to be *congruent modulo m* if their difference $x - y$ is exactly divisible by $m$. For example, 5 and 13 are congruent modulo 4, because 4 divides $-8$, which is $5 - 13$. Similarly, 38 and 10 are congruent modulo 7, because 7 divides 28. But 15 and 8 are not congruent modulo 3.

When a positive integer $a$ is divided by a positive integer $m$, we obtain a quotient $q$ and a remainder $r$:

$$a = qm + r$$

The remainder satisfies the inequality

$$0 \leqslant r < m$$

So another way of saying that $x$ and $y$ are congruent modulo $m$ is to say that $x$ and $y$ both leave the same remainder when divided by $m$; it being understood that 'remainder' always refers to that unique value $r$ satisfying the above inequalities.

The notation for '$x$ and $y$ are congruent modulo $m$' is

$$x \equiv y \; (\text{mod } m)$$

Alternatively, we say that '$x$ is congruent to $y$ modulo $m$'. Congruences play an important part in a branch of mathematics called number theory, and the application of this branch plays an important part in cryptography.

If

$$x \equiv y \; (\text{mod } m) \quad \text{and} \quad u \equiv v \; (\text{mod } m)$$

then it is easy to show that

$$x + u \equiv y + v \; (\text{mod } m)$$

$$x - u \equiv y - v \; (\text{mod } m)$$

$$xu \equiv yv \; (\text{mod } m)$$

Essentially, this means that the remainders obey the same arithmetic laws of addition, subtraction and multiplication as do the ordinary integers. There are, of course, $m$ different possible remainders when dividing by $m$. These are the integers $0, 1, 2, \ldots, m - 1$, and are called the *residues modulo m*. Two residues can be added to produce a third, as follows: they are added in the normal manner and the answer is divided by $m$ to obtain the remainder; the sum of the two residues is then that remainder. As an example, we take $m = 5$ and consider the sum of the two residues 3 and 4. Now, 3 plus 4 equals 7, and 7 divided by 5 leaves the remainder, 2. Hence, the sum of the two residues 3 and 4 is 2. We need a notation to express all this, rather than use words. We write, simply,

$$3 + 4 \equiv 2 \ (\mathrm{mod} \ 5)$$

making use of our previous notation.

We can make a table of the result of adding two residues modulo $m$. For $m = 5$ it is

| + | 0 | 1 | 2 | 3 | 4 |
|---|---|---|---|---|---|
| 0 | 0 | 1 | 2 | 3 | 4 |
| 1 | 1 | 2 | 3 | 4 | 0 |
| 2 | 2 | 3 | 4 | 0 | 1 |
| 3 | 3 | 4 | 0 | 1 | 2 |
| 4 | 4 | 0 | 1 | 2 | 3 |

Similar rules hold in 'modular arithmetic' for subtraction and multiplication. For example, the residue 3 subtracted from the residue 1 modulo 5 is calculated as follows: $1 - 3$ equals $-2$, and

$$-2 \equiv 3 \ (\mathrm{mod} \ 5)$$

so the difference is the residue 3. The product of the two residues 3 and 4 modulo 5 is the residue 2, because

$$3 \times 4 \equiv 2 \ (\mathrm{mod} \ 5)$$

The tables for subtraction and multiplication modulo $m$ can also be constructed. For $m = 5$, they are

| − | 0 | 1 | 2 | 3 | 4 |
|---|---|---|---|---|---|
| 0 | 0 | 4 | 3 | 2 | 1 |
| 1 | 1 | 0 | 4 | 3 | 2 |
| 2 | 2 | 1 | 0 | 4 | 3 |
| 3 | 3 | 2 | 1 | 0 | 4 |
| 4 | 4 | 3 | 2 | 1 | 0 |

| × | 0 | 1 | 2 | 3 | 4 |
|---|---|---|---|---|---|
| 0 | 0 | 0 | 0 | 0 | 0 |
| 1 | 0 | 1 | 2 | 3 | 4 |
| 2 | 0 | 2 | 4 | 1 | 3 |
| 3 | 0 | 3 | 1 | 4 | 2 |
| 4 | 0 | 4 | 3 | 2 | 1 |

Let's return to Sam and Boris for a while, because they now know enough mathematics to produce a new block cipher. Whether or not it is any good will depend on the ingenuity of Maxine, so perhaps they will need to modify it later!

## *Scene 1*

Sam and Boris are having a secret meeting to discuss how they are going to use their new-found mathematics to thwart the MI5 agent Maxine. They have discarded their Caesar cipher and permutation cipher. They are looking for something more sophisticated. They have learnt some mathematics and they both know how to program their home microcomputers.

BORIS: Sam, as I see it, our first task when sending messages is to code all the letters of the alphabet and to code such characters as space, comma, full stop and question mark. If we code them into binary digits or bits, how many bits do we need for each symbol?

SAM: Let's see. There are 26 letters – we needn't bother about distinguishing between upper- and lower-case letters – plus 4 more characters. That's 30 characters in all. So we'll need at least 5 bits, since $30 < 32 = 2^5$. We could use the binary representation of the numbers from 1 to 26 for the letters A to Z, 27 to 30 for the others. Our coding table would then look like this:

| A | 00001 | G | 00111 | M | 01101 | S | 10011 | Y | 11001 |
|---|---|---|---|---|---|---|---|---|---|
| B | 00010 | H | 01000 | N | 01110 | T | 10100 | Z | 11010 |
| C | 00011 | I | 01001 | O | 01111 | U | 10101 | – | 11011 |
| D | 00100 | J | 01010 | P | 10000 | V | 10110 | , | 11100 |
| E | 00101 | K | 01011 | Q | 10001 | W | 10111 | . | 11101 |
| F | 00110 | L | 01100 | R | 10010 | X | 11000 | ? | 11110 |

BORIS: That's good, Sam. We have two representations left over – 00000 and 11111 – if we want to introduce any further characters. Now you can write a message:

10011101000000101110110110011101111011000101

SAM: Yes, and when you decode, all you have to do is divide the code up into groups of 5 bits and substitute the right letter:

10011/10100/00001/01110/11011/00111/01111/01110/00101

    S     T     A     N     –     G     O     N     E

Boris, I don't think it will take Maxine very long to work this out. It is a bit like our substitution cipher, and you know how quickly she cracked that. We only used it once and poor Stan was nabbed!

BORIS: The *coding* into bits is only the first stage, Sam: the next stage is the *enciphering*, where we use our modular arithmetic. You take 8 bits at a time and get a number $p$ between 0 and 256. Actually, to be more precise, $0 \leqslant p < 256$.

SAM: Where did you dream up that number 256 from? Oh, I know, $256 = 2^8$.

BORIS: Quite so, Sam. You latch on quickly! Now comes the enciphering part. You and I agree a key, $k$. It will have to be an odd number between 0 and 256. Suppose today it is 73. You, the encipherer, multiply the plaintext $p$ by the number $k$, divide the result by 256 and take the remainder to be ciphertext. This will give you a number $c$ which satisfies $0 \leqslant c < 256$. In modular arithmetic notation, this is the same as saying $c \equiv pk \pmod{256}$. Let's see, the first 8 bits of our message are 10011101, which is binary for the number 157. Now, $157 \times 73 = 11\,461$, and $11\,461 = (44 \times 256) + 197$. So our ciphertext is 197, which in binary is, if I'm not mistaken, 11000101. So these are the first 8 bits of the ciphertext. You must do the same for the next 8 bits, and so on.

SAM: It's a good job we've got our computers to do the arithmetic. If I had to do it by hand I'd make all sorts of mistakes! Boris, that message above has 45 bits. That makes five groups of 8 bits, and 5 bits left over. How do I encipher them?

BORIS: You can pad the message out by adding extra bits so that after encoding there is always an exact multiple of 8 bits. In this case you need 3 extra bits. They can be anything. When the ciphertext is received and deciphered, the extra bits can be ignored. Let's see, if you pad the last 5 bits with 111 you'll get 00101111 for the last block. To encipher it proceed as follows:

$$00101111 = 47 \quad \text{in decimal}$$

$$47 \times 73 \equiv 103 \pmod{256}$$

$$103 = 01100111 \quad \text{in binary}$$

SAM: It looks simple enough, Boris. But you haven't told me how you are going to decipher the ciphertext. You can't divide 103 by 73 to get 47, can you?

BORIS: Well, yes, you can in a way. I'll need to solve $73 \times p \equiv 103 \pmod{256}$ to find $p$. This isn't too bad, because I know that $249 \times 73 \equiv 1 \pmod{256}$. So if I multiply the equation in $p$ by 249 on both sides, I get $249 \times 73 \times p \equiv 249 \times 103 \pmod{256}$, or $1 \times p \equiv 47 \pmod{256}$.

SAM: That is ingenious, Boris! So I encipher my plaintext 8 bit number by multiplying by 73 (mod 256), and you decipher by multiplying by 249 (mod 256).

BORIS: That's right. Simple, isn't it? We have two keys: an *enciphering key*, which in this case is 73, and a *deciphering key*, which in this case is 249.

SAM: Ah, but Boris, you said my enciphering key could be any odd number and might change from week to week. How will these changes affect your deciphering key?

BORIS: You have to tell me what you are changing it to, and then I can work out the new deciphering key. If you choose $k$, I have to find $d$ such that $kd \equiv 1 \pmod{256}$. This can always be solved if $k$ is an odd number and the method of solution is quite practical and easy.

*Interlude*

We interrupt Boris and Sam at this moment to make a few comments about their new-found cipher. What they have devised is an example of a *linear block cipher*. The choice of 8 bits for the block length, leading to a modulus of 256, was arbitrary. A small block length makes the arithmetic easier, but on a computer, and with special procedures for dealing with large numbers, it wouldn't be difficult to have blocks of 512 bits or more. With a larger block of $b$ bits, the modulus would be $2^b$, with a larger choice of enciphering keys. Since any odd number between 0 and $2^b$ is a possible key, there is a choice from $2^{b-1}$.

The problem we have yet to consider about such a block cipher is the determination of the deciphering key $d$ from the enciphering key $k$. As Boris pointed out, we have to solve

$$kd \equiv 1 \pmod{2^b}$$

This is not as serious as it seems – even though trial and error is out of the question. The method is actually very old and goes back at least to the Greek mathematician Euclid. The method, known as *Euclid's algorithm*, determines $d$ very quickly.

To illustrate the method, we show how Boris deduced the deciphering key for the enciphering key of 73 and the modulus 256. The first step is to divide 73 into 256, to get a quotient 3 and remainder 37:

$$256 = 3 \times 73 + 37$$

The next step is to take the remainder 37 and divide it into 73 to get a quotient 1 and remainder 36:

$$73 = (1 \times 37) + 36$$

These steps are repeated again and again. You take the divisor of the previous step, and divide it by the remainder of the same step to get a new quotient and remainder. The remainders are going down all the time, and in general quite rapidly. Eventually, the remainder will be 0. The remainder of the previous step is the highest common factor of the two numbers we started with. In Boris's case, this happens soon:

$$37 = (1 \times 36) + 1$$
$$36 = (36 \times 1) + 0$$

So 1 is the highest common factor of 73 and 256. We knew this already, but the calculations will help us to solve our equation. Using the penultimate step, we have 1 in terms of 37 and 36:

$$1 = 37 - (1 \times 36)$$

Using the step before, we can express 1 in terms of 73 and 37 by replacing 36 by $73 - (1 \times 37)$. Thus

$$1 = 37 - \{1 \times [73 - (1 \times 37)]\}$$
$$= (2 \times 37) - 73$$

Using the step before, we replace 37 by $256 - (3 \times 73)$ to get

$$1 = \{2 \times [256 - (3 \times 73)]\} - 73$$
$$= (2 \times 256) - (7 \times 73)$$

This last expression, put another way, says that $-7 \times 73 \equiv 1$ (mod 256), and since we prefer to deal in positive numbers, and

$$-7 \equiv 249 \ (\text{mod } 256)$$

we get the solution $d \equiv 249$ that Boris claimed.

Let's look at another key and modulus, to see what happens when the numbers are bigger. Suppose we want to solve

$$387d \equiv 1 \ (\text{mod } 4096)$$

The arithmetic proceeds as follows:

$$4096 = (10 \times 387) + 226$$
$$387 = (1 \times 226) + 161$$
$$226 = (1 \times 161) + 65$$
$$161 = (2 \times 65) + 31$$
$$65 = (2 \times 31) + 3$$
$$31 = (10 \times 3) + 1$$
$$3 = (3 \times 1) + 0$$

Hence

$$1 = 31 - (10 \times 3)$$
$$= 31 - \{10 \times [65 - (2 \times 31)]\}$$
$$= (21 \times 31) - (10 \times 65)$$
$$= \{21 \times [161 - (2 \times 65)]\} - (10 \times 65)$$
$$= (21 \times 161) - (52 \times 65)$$
$$= (21 \times 161) - [52 \times (226 - 161)]$$
$$= (73 \times 161) - (52 \times 226)$$
$$= 73 \times [387 - (1 \times 226)] - (52 \times 226)$$
$$= (73 \times 387) - (125 \times 226)$$
$$= (73 \times 387) - \{125 \times [4096 - (10 \times 387)]\}$$
$$= (1323 \times 387) - (125 \times 4096)$$

Hence $d = 1323$ is the required deciphering key.

### Scene 2 (a few months later)

Boris and Sam are having another secret meeting to discuss their linear block cipher.

BORIS: Well, Sam, we should be pleased with ourselves. We seem to have that MI5 agent licked now. I don't think she has been able to decipher our ciphertext recently.

SAM: Boris, I hope you are right. It wasn't good at the beginning, though. First, she worked out our code for changing characters to

5 bit numbers, then she guessed we were using an 8 bit linear block cipher. It didn't take her long to try all possible 128 odd numbers to discover the right deciphering key when the message made sense.

BORIS: Yes, but meanwhile we've lost quite a few agents. Whatever happened to Jane in Accrington?

SAM: A very unfortunate accident, that! Still, when we increased the block size it got better, didn't it?

BORIS: It certainly did, Sam. But I had to warn you that our cipher wouldn't be very secure if Maxine knew a plaintext and corresponding ciphertext block. When she knew $p$ (plaintext) and $c$ (ciphertext) and the modulus $2^b$, she had only to solve $cd \equiv p \pmod{2^b}$, by Euclid's algorithm, to get $d$. Then she could decipher the rest of the message easily. So, Sam, that's why you must be careful not to begin or end your messages with anything obvious like your address or 'Yours ever, Sam'.

SAM: OK, I get the message.

BORIS: And so did Maxine! We still have problems, Sam. Our block size is large, so our enciphering keys are very long and difficult to remember. That book I gave you with the list of keys – one for each week of the year. I am afraid it will be discovered by Maxine. Your house is not physically secure like my embassy. What I would really like is a cipher for which, even if Maxine discovered your enciphering key, she still wouldn't be able to decipher your messages.

SAM: Does that mean we need to learn some more mathematics?

BORIS: I'm afraid so, Sam.

From the above scene you can see that the linear block cipher based on multiplication by the modulo $m$ has serious disadvantages. The deciphering key can easily be discovered – from a block of known plaintext and ciphertext. The enciphering key is long, and requires to be kept physically secure since the cryptanalyst could easily deduce the deciphering key from it.

The cipher has other deficiencies, too. So Sam and Boris were certainly misguided in having any confidence in it. Maxine will surely have noticed the following. Suppose the plaintext were the number $2^{b-1}$ and the enciphering key is any odd number $k = 2k' + 1$. Then

$$pk = 2^{b-1}k = 2^b k' + 2^{b-1} \equiv 2^{b-1} \pmod{2^b}$$

This shows that the corresponding ciphertext is also $2^{b-1}$, no matter what the key is. More generally, plaintext of the form $2^f$ will have ciphertext also divisible by $2^f$, which is certainly undesirable.

The cipher also has weak keys where plaintext and corresponding ciphertext are always closely related. You might like to ponder why $k = 2^{b-1} + 1$ is a weak enciphering key.

## Data Encryption Standard

Good block ciphers are not linear, because too much mathematical structure allows the cryptanalyst to exploit mathematical ideas for breaking them. In practice, block ciphers use a mixture of techniques employing structure and non-structure. The non-structured part is often effected by permuting the bits of the intermediate results. On p. 252 we stated some of the properties of a good cipher. Let's expand on these. A cipher should have the following properties:

(1) Cheap to implement. This means that the equipment necessary to encipher and decipher doesn't cost more than the information it is trying to secure.
(2) Speedy. The algorithms for enciphering and deciphering must keep pace with the rate at which the information is to be transmitted.
(3) A large number of possible keys; large enough to ensure that the cryptanalyst cannot reasonably try them all.
(4) Difficult to determine the keys from some known plaintext and corresponding ciphertext. Here 'difficult' means either taking an unreasonable amount of time, or requiring resources (e.g. hundreds of computers) beyond the value of the information.
(5) Statistical tests applied to the ciphertext must not reveal either complete or partial information about the plaintext.
(6) A small change in the block of plaintext results in each individual bit of the ciphertext changing with probability about one-half. This prevents the cryptanalyst gaining information from similar blocks of ciphertext.

In 1977, the cipher known as the Data Encryption Standard (DES) was published in the USA. This cipher is now used widely throughout the world. Special computer chips to implement it have been manufactured, and they can encipher and decipher about 20 million bits per second. The key is essentially a 56 bit number, and the cipher enciphers blocks of 64 bits at a time. The cipher is too complex to describe here, but has the favourable properties indicated above. There is, however, some controversy over the DES cipher. Some experts think that a 56 bit key is not big enough; others are worried that the cipher may have been designed in such a way that the designers know a special secret method of breaking it. Such a method is known as a *trapdoor* attack. Normal users of the cipher believe it to be secure, and cannot see the trapdoor, but any cryptanalyst who might be in league with the designers could open the trapdoor and discover plaintext easily.

## Exponentiation ciphers

Earlier, we considered ciphers based on multiplying plaintext by a key $k$ modulo $m$. *Exponentiation ciphers*, which involve raising to a power, work in a similar way. In this case, however, if $m$ is the modulus, $p$ the plaintext (so $0 \leqslant p \leqslant m - 1$) and $e$ the enciphering key, the ciphertext $c$ is given by

$$0 \leqslant c \leqslant m - 1 \quad \text{and} \quad c \equiv p^e \text{ (mod } m)$$

In practice $m$ and $e$ are going to be, perhaps, 512 bit numbers, numbers with something like 150 decimal digits. If we were to compute $p^e$ as an integer, it might have $10^{150}$ digits or more, so there had better be an easier way of computing $c$. Fortunately, there is.

First of all, whenever we do an intermediate multiplication to obtain $c$, we can also reduce the answer modulo $m$ to obtain a number between 0 and $m - 1$, and work from this. Thus, to obtain $c \equiv 3^7$ (mod 17), instead of evaluating $3^7 = 2187$ and dividing by 17 to get the remainder $c = 11$, we could proceed as follows:*

$$3^2 \equiv 3 \times 3 \equiv 9 \text{ (mod 17)}$$

$$3^3 \equiv 3 \times 3^2 \equiv 3 \times 9 \equiv 27 \equiv 10 \text{ (mod 17)}$$

$$3^4 \equiv 3 \times 3^3 \equiv 3 \times 10 \equiv 30 \equiv 13 \text{ (mod 17)}$$

$$3^5 \equiv 3 \times 3^4 \equiv 3 \times 13 \equiv 39 \equiv 5 \text{ (mod 17)}$$

$$3^6 \equiv 3 \times 3^5 \equiv 3 \times 5 \equiv 15 \text{ (mod 17)}$$

$$3^7 \equiv 3 \times 3^6 \equiv 3 \times 15 \equiv 45 \equiv 11 \text{ (mod 17)}$$

The answers are the same; the second method avoids our ever having to multiply numbers bigger than $m$.

When the exponent $e$ is large, there is another technique which can save time and effort when computing $c \equiv p^e$ (mod $m$). To illustrate it, let's look at an example: say we want to compute

$$c \equiv 3^{13} \text{ (mod 17)}$$

---

* In this notation, any quantity in a given line and any other quantity in that line are congruent mod 17.

We can calculate $3^2$ (mod 17), $3^4$ (mod 17) and $3^8$ (mod 17) very easily, just by successive squaring and taking the remainder mod 17. Thus

$$3^2 \equiv 9 \text{ (mod 17)}$$
$$3^4 \equiv (3^2)^2 \equiv 9^2 \equiv 81 \equiv 13 \text{ (mod 17)}$$
$$3^8 \equiv (3^4)^2 \equiv 13^2 \equiv 169 \equiv 16 \text{ (mod 17)}$$

And so

$$3^{13} \equiv 3^{8+4+1} \equiv 3^8 \times 3^4 \times 3^1$$
$$\equiv 16 \times 13 \times 3 \text{ (mod 17)}$$
$$\equiv 16 \times 39 \equiv 16 \times 5 \equiv 80 \equiv 12 \text{ (mod 17)}$$

Thus $c = 12$. We calculated it by doing five multiplications of numbers less than 17, which is certainly an improvement over the twelve multiplications required by successive multiplications by 3. You might like to check the answer on your calculator or computer. Then you might like to have a go at working out $c$ when $p = 7$, $e = 19$ and $m = 23$.

The techniques described above for calculating $p^e$ (mod $m$) are even more beneficial when $e$ and $m$ are 512 bit numbers. To compute $c$ requires about a thousand multiplications of 512 bit numbers, a lot simpler than the original calculation we considered. Mathematical ideas have come to the rescue!

So far we have discussed how to encipher plaintext $p$ and obtain ciphertext $c$ using the exponentiation cipher. How is the decipherer going to deduce $p$ from the values of $c$, $m$ and $e$? Logarithms are no use. It turns out that the deciphering algorithm is going to be exactly the same, but that the *key* is different. Thus the decipherer has an exponent $d$ and computes $p$ from

$$p \equiv c^d \text{ (mod } m)$$

I don't need to tell you about how the decipherer computes $p$ – the techniques are those used by the encipherer. But I *do* need to tell you how the number $d$ is related to $e$ and $m$. For the number $d$ to work, it must be true that, for *any* plaintext $p$,

$$\text{if} \quad c \equiv p^e \text{ (mod } m) \quad \text{then} \quad p \equiv c^d \text{ (mod } m)$$

This means that for any $p$

$$p \equiv (p^e)^d \text{ (mod } m)$$

or, equivalently,

$$p \equiv p^{ed} \pmod{m}$$

That there is such a $d$ which works for all $p$ is a consequence of the mathematical structure of the set of residues modulo $p$.

We need to find for which values $x$ it is true that for all $p$

$$p \equiv p^x \pmod{m}$$

Some fairly elementary results in number theory come to our rescue again. (It seems more than a coincidence that 'secure' and 'rescue' are anagrams!) It can be shown that, for any number $m$, there is a corresponding number $t(m)$ which has the following property. Suppose $a$ is *any* number, whose only factor in common with $m$ is the factor 1; $a$ is said to be *coprime* to $m$, and mathematicians write it in shorthand as $(a, m) = 1$. Then

$$a^{t(m)} \equiv 1 \pmod{m}$$

For example, if $m = 7$, then $t(m)$ has the value 6 and

$$1^6 \equiv 2^6 \equiv 3^6 \equiv 4^6 \equiv 5^6 \equiv 6^6 \equiv 1 \pmod{7}$$

If $m = 15$, then $t(m)$ has the value 4 and

$$1^4 \equiv 2^4 \equiv 4^4 \equiv 7^4 \equiv 8^4 \equiv 11^4 \equiv 13^4 \equiv 14^4 \equiv 1 \pmod{15}$$

The value of $t(m)$ is not difficult to find if you know the factorization of $m$. Thus Pierre de Fermat knew in the seventeenth century that, if $m$ is a prime number, then $t(m) = m - 1$, and if $m$ is the product of two different primes, $m = q_1 q_2$, then $t(m)$ is the lowest common multiple of $q_1 - 1$ and $q_2 - 1$. For example, the lowest common multiple of $5 - 1$ and $7 - 1$ is 12, so $t(35) = 12$.

Returning to our original problem, we wanted for a given $m$ the numbers $x$ such that

$$p \equiv p^x \pmod{m}$$

Now, if $(p, m) = 1$, then

$$p^{t(m)+1} \equiv p^{t(m)} \times p \equiv p \pmod{m}$$

So $x = t(m) + 1$ is one possibility in this case. But for every number $n$,

$$p^{nt(m)+1} \equiv (p^{t(m)})^n \times p \equiv 1^n \times p \equiv p \pmod{m}$$

so any $x$ of the form $nt(m) + 1$ is also a possibility.

So when our decipherer needs to determine $d$ from $e$ and $m$, so that he can decipher, all he has to do is calculate $t(m)$, then solve $ed \equiv 1$ (mod $t(m)$). The first requires knowing the factorization of $m$; the second can be done by using Euclid's algorithm, and also it requires the exponents $e$ to be restricted to those that satisfy

$$(e, t(m)) = 1$$

Let's see how Boris would decipher if Sam had sent him a message using the exponentiation cipher. Suppose Sam's plaintext were the number 3, and he was using the modulus 17 and the exponent 13. We have already seen that

$$12 \equiv 3^{13} \text{ (mod 17)}$$

so that the ciphertext is $c = 12$.

Boris first has to compute the right value for $d$. He doesn't do this every time, of course; once $e$ and $m$ are fixed, he calculates $d$, memorizes it, and then uses it every time to get the plaintext from the ciphertext. Since 17 is prime, $t(17) = 16$ and Boris has to solve

$$13d \equiv 1 \text{ (mod 16)}$$

Now, $16 = 13 + 3$ and $13 = 4 \times 3 + 1$, so

$$1 = 13 - (4 \times 3) = 13 - [4 \times (16 - 13)]$$
$$= (5 \times 13) - (4 \times 16)$$

Hence $d = 5$. When Boris receives the ciphertext 12, he computes $p = 12^5$ (mod 17) as follows:

$$12^2 \equiv 144 \equiv 8 \text{ (mod 17)}$$
$$12^4 \equiv 8^2 \equiv 64 \equiv 13 \text{ (mod 17)}$$
$$12^5 \equiv 13 \times 12 \equiv 156 \equiv 3 \text{ (mod 17)}$$

Boris gets the plaintext $p = 3$, which naturally agrees with Sam's message.

In fact, this cipher, as it stands, is only a little better than Sam and Boris's previous cipher based on multiplication modulo $m$. It is a little more secure, in the sense that it is harder to deduce $e$ or $d$ from known plaintext and ciphertext, and it is less susceptible to weak plaintext and weak keys.

The cipher does allow us, however, to explain a new cipher which is practicable and has one remarkable property.

## Public key cryptography

With the ciphers we have looked at so far, for anyone knowing the enciphering algorithm and key it would not be difficult to determine the deciphering algorithm and key. Until 1978, all ciphers were of that form. They required a certain amount of trust between the encipherer and decipherer. How is Boris to know that Sam is not a double agent, revealing his enciphering key to Maxine, but pretending to Boris that the messages are secure? Another disadvantage, as we have seen, is that enciphering keys have to be physically secure.

Is it possible to have a cipher where everybody knows the enciphering algorithm *and the enciphering key*, yet only the decipherer can decipher messages? Think what an advantage that would be to Boris. All his agents (Sam, Stan, even Jane) could use this cipher to send him messages. None of them could read the messages of other agents because they would not know how to decipher. Similarly, Maxine would reap no benefit by discovering the enciphering key. She can only decipher when she knows Boris's deciphering key, and that is securely locked away in his embassy.

### Public key cipher

A public key cipher is the exponentiation cipher (p. 263), but with a subtle difference. The decipherer decides on two very large prime numbers, $q_1$ and $q_2$. He may do this, for example, by writing down two 256 bit random numbers, $n_1$ and $n_2$, and choosing $q_1$ and $q_2$ to be the smallest primes greater than $n_1$ and $n_2$ respectively. There are quick computer tests to check whether or not a number $q$ is prime – even numbers with a hundred decimal digits. If $q$ is not prime, however, the test only reveals that; it doesn't actually tell you a factorization of $q$.

The decipherer then multiplies his primes $q_1$ and $q_2$ together to obtain his modulus $m = q_1 q_2$. Since he knows $q_1$ and $q_2$, he can obtain $t(m)$ as the lowest common multiple of $q_1 - 1$ and $q_2 - 1$. He then chooses a random number $e$ such that $(e, t(m)) = 1$, and from $e$ and $t(m)$ he easily computes $d$, as already described.

Now, all the ingredients of the cipher are there. He publishes openly the numbers $e$ and $m$, but keeps secret the values of $d$, $q_1$, $q_2$ and $t(m)$. In fact, he can throw away the values of $q_1$, $q_2$ and $t(m)$, if he wants; he is only going to use the value of $d$.

He invites anyone to send him secret messages, enciphering $p$, as before, by the formula $c \equiv p^e \pmod{m}$. He deciphers such ciphertext by the formula $p \equiv c^d \pmod{m}$, using the secret value of $d$.

If Boris were to use this public key cipher, giving Sam the values of $e$ and $m$, how can he be sure that Maxine, knowing $e$ and $m$, won't be able to determine $d$, and so then be able to decipher ciphertext $c$? Recall how $d$ was calculated. If Maxine is to repeat this calculation she will need to know $t(m)$. This depends crucially on those prime factors $q_1$ and $q_2$ of $m$. Boris began with $q_1$ and $q_2$, and then calculated $m$. But Maxine will have to deduce $q_1$ and $q_2$ from the value of $m$. For 512 bit numbers $m$ which are the product of two 256 bit random primes, this problem turns out to be no easy matter. Earlier, we said that checking whether or not a number is prime is relatively easy, but the method only gives the answer 'Yes, it is' or 'No, it isn't'; in the latter case it doesn't tell you the factors. With the present state of mathematical knowledge, the problem of finding the factors is a very difficult one indeed. To factorize a 256 bit number which is the product of two 128 (or so) bit primes is just about feasible on a large computer, and with a bit of luck, in an hour. (The technique is very sophisticated and cannot be explained here; trial division by 2 and odd numbers 3, 5, 7, . . . is not suitable.) Using the same sophisticated techniques and fast computer to factorize a 512 bit number would take years. The secret would have to be very important for it to be worth waiting that long.

You may have thought that factorizing a number into primes was a topic of pure mathematics, a puzzle, an abstract problem without practical use. Cryptography has now turned this topic into one of applied mathematics, and mathematicians are now trying to understand and to quantify how difficult this problem of factorization is.

If you think this is the end of the story, well it isn't. Public key ciphers are now used extensively, but since they are not so economical they are often used only for special purposes, for example when two parties need to exchange a DES key or when the receiver of a message needs to validate the identity of the sender.

Where does this leave our spies Boris and Sam, and our MI5 agent Maxine? Right now, cryptography seems to have given Boris the upper hand: securing a message seems to be feasible. But that doesn't mean that Maxine has given up. She is studying mathematics and hopes to come up with a new idea. Anyway, perhaps we shouldn't be too unhappy about Boris's success. At the level where cryptography affects you and me, it is important that data relevant to us, whether medical, financial or whatever, can be transmitted securely.

## Summary

In this chapter we have looked at various ciphers. In particular, we saw how mathematics has played a part in the development of modern secure systems where the rate of information transfer is perhaps 20 million bits per second. We have also looked at the public key ciphers.

Good ciphers need to be cheap, fast and, of course, very secure. A deciphering algorithm should be no more complex than the corresponding enciphering algorithm. Mathematics has helped in the design of ciphers having these properties. But, curiously, mathematics has played a negative role in two ways. First, some ciphers are no good because they are based entirely on mathematical structures and are therefore open to attack by methods which exploit this structure. These ciphers have to be modified so that they are less structured, but also more secure. Second, some ciphers are secure because mathematics cannot provide an efficient method of factorizing a very large number. There cannot be many topics which prosper on the shortcomings of mathematics!

## Further reading

D. Welsh, *Codes and Cryptography* (Oxford Science Publications, 1988).

# 12

## SUPERCOMPUTERS:
## HOW SUPER ARE THEY?

DAVID JACOBS

### One man went to dig

In years past, part of the process of learning the basic rules of arithmetic was answering questions such as

Q: If it takes one man one day to dig a hole, how long will it take two men to dig an identical hole?
A: Obvious – half a day.

But is it? What if the hole is such a size and shape that only one man can work in it at a time? If it is, then allocating two men to the task will not speed up the process. Thus there are some jobs that cannot be speeded up simply by allocating more effort.

But let's look at the whole job. With the hole dug, perhaps the next task is to mend an electric cable, then fill the hole in and finally resurface the pavement (see Fig. 12.1). Say each job takes a day. So that is four days for one man.

Q: How long will it take four men?
A: Again, the answer is four days, if the jobs have to be done sequentially, and each can be done only by one man at a time.
Q: But what if we have four cables to mend?
A: Then with one man it will take 16 days, but with four men it could be done in four days.

But that assumes each man can do all four jobs (equally well).

Suppose we have one man who can dig holes, a different man who can mend cables, another who can fill holes, and a fourth who can surface pavements (see Fig. 12.2).

Q: How long will it take the four specialists to mend one cable?
A: Four days, during which time each man only works for one day.

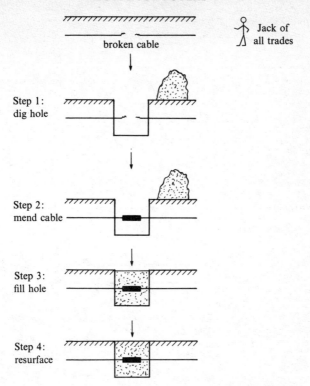

*Figure 12.1.* An example of a SISD (Single Instruction, Single Datastream) processor: a 'Jack of all trades'.

But if we have four cables to mend . . .

Q: How long will it take the four specialists to mend the four cables?

A: Not 16 days, but 7.

The tasks done each day are listed in Table 12.1. If there is a never-ending supply of broken cables to be mended, then the four-man team can start a new cable repair each day, and complete one per day, having four repairs in progress in parallel at any one time.

The measurement of performance of computers, particularly supercomputers, has similarities with the measurement of the performance of our cable-mending team. Supercomputers tend to make use of *parallelism* ('many hands make light work') – equivalent to using our four universal men for tackling the cable-mending tasks above, and exploit *replication* of the problem whereby each job or task has to be done many times over, but at different places (e.g. on

broken
cable

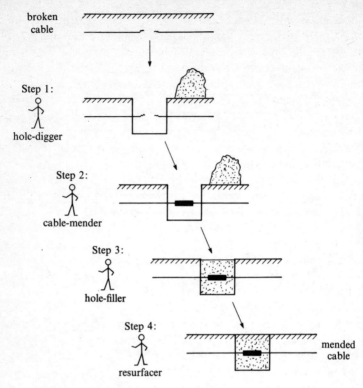

Step 1:
hole-digger

Step 2:
cable-mender

Step 3:
hole-filler

Step 4:
resurfacer

mended
cable

*Figure 12.2.* An example of a pipeline process with a team of
four, each working on a specialist task to produce a
through-flow of output.

*Table 12.1.* Order of tasks done by specialist workmen.

| Day | Hole-digger | Cable-mender | Hole-filler | Resurfacer |
|-----|-------------|--------------|-------------|------------|
| 1 | Cable 1 | | | |
| 2 | Cable 2 | Cable 1 | | |
| 3 | Cable 3 | Cable 2 | Cable 1 | |
| 4 | Cable 4 | Cable 3 | Cable 2 | Cable 1 |
| 5 | | Cable 4 | Cable 3 | Cable 2 |
| 6 | | | Cable 4 | Cable 3 |
| 7 | | | | Cable 4 |

different data). Indeed, with computers able to process many millions of numbers a second, supercomputing power is generally needed only to solve problems which feature large vectors and arrays – programs not having many repetitive operations tend to require only modest computing time. Many supercomputers also use *pipelining*, whereby each job is subdivided into separate subtasks, and (specialist) units deal with each subtask in turn, the units being able to work in parallel. This is equivalent to our four specialists working at cable-mending, each doing his own specialist job, but cooperating to produce work throughput.

But how do we measure the performance of such computers? A simple empirical way would be just to run them on test problems, and compare times. But this may hide important features of the machines and their performance traits. Such simple empirical comparisons are little help when it comes to designing computers or writing computer programs, and understanding why some machines perform better on some problems than others. We need a model which encapsulates the key features.

Nowadays computers are commonplace, not only in industry and business, but also in schools and in the home. These computers are most often used to manipulate numbers – namely, to do mathematics – which, of course, they can do at great speed. So they are very good at doing mathematics, but is there anything that mathematics can do for them? Well of course there is. Mathematics is an integral part of the methods used to specify, design and manufacture computers. Let us pick just one of these areas, one which at first seems too obvious to need further attention: How fast are computers? And, in particular, how does one compare the performance of one computer with another? Here, as with so many other fields of measurement, mathematics contributes to many of the answers.

### Straightforward sequential computers

Comparisons of the performance of different computers were relatively straightforward when the basic architecture of 'nearly all computers' was virtually identical, as it was until about the mid-1960s. Even today the majority of computers are very similar. They are basically sequential processors, capable of one operation at a time, and operations are performed in the order that is called for by the program. However, there are computers which are designed to have a performance that far outstrips that of normal computers, and such computers are often called *supercomputers*. In order to achieve the performance levels required from such machines, it often proves inadequate just to use the fastest logic components to make the computer. So new designs are developed, and it is with these designs that comparisons then become difficult, and require more than simple measures of the number of operations performed in one second.

With a sequential processor, measurement is relatively easy. The processor will be run at a particular clock rate, which is the speed at which pulses are generated, during each of which the computer is capable of performing one instruction or simple operation. Thus sequential computers are rated at so many instructions per second, or 'ips', and the powerful computers of today are rated in 'mips', millions of instructions per second. Thus a computer with a 1000 Hz clock rate would typically be rated at 1000 ips. A more widely used term for the computer hardware is the *cycle time*, or clock period, the time taken for an instruction to be executed. For a straightforward sequential computer, if the cycle time is $\tau$ seconds, the number of instructions performed per second is $1/\tau$. Thus a 50 nanosecond cycle time (50 ns, or $50 \times 10^{-9}$ seconds) would correspond to a 20 mips performance.

Many of the most demanding calculations for which the fastest computers are needed consist of arithmetic operations on numbers which are not integers, and are expressed as decimal fractions. We call these *floating-point numbers*, because the position of the leading digit relative to the decimal point may be different for the different numbers in the calculation, so the decimal point is 'floating'. Such numbers are effectively stored in the computer in compact form. For example,

$$3.1416 \quad \text{will be stored as} \quad 0.314\,16 \times 10^1$$

$$314.16 \quad \text{will be stored as} \quad 0.314\,16 \times 10^3$$

$$-0.031\,416 \quad \text{will be stored as} \quad -0.314\,16 \times 10^{-1}$$

Arithmetic operations on such numbers usually require more work, or computation, than the same operations performed on integers (or *fixed-point numbers*). Thus, the addition of two floating-point numbers requires

(1) the comparison of the exponents
(2) the shifting of the decimal point in the smaller number by the difference in the exponents

*Table 12.2.* Addition of $3.1416 \times 10^2$ and $6.85 \times 10^{-1}$, a typical subdivision of the simple arithmetic operation of adding two floating-point numbers.

| Step 1 | Compare exponents | $2 - (-1) = 3$ |
|---|---|---|
| Step 2 | Shift decimal point of 6.85 by 3 | 0.006 85 |
| Step 3 | Add 3.1416 to 0.006 85 | 3.148 45 |
| Step 4 | Normalize the result | $3.148\,45 \times 10^2$ |

(3) the addition of the two normalized numbers
(4) the normalization of the answer to the standard format.

For example, consider adding $3.1416 \times 10^2$ and $6.85 \times 10^{-1}$. The first number is larger in magnitude, and the difference in exponents is 3, so we adjust the smaller number so that we now have to add $3.1416 \times 10^2$ and $0.00685 \times 10^2$ (see Table 12.2).

For such operations the term 'flops', meaning *floating-point operations per second*, is used, with Mflops, or megaflops, meaning $10^6$ flops, and Gflops, or gigaflops, meaning $10^9$ flops.

It is the measurement of performance on computers other than straightforward sequential computers to which we now turn our attention – bearing in mind the examples of men digging holes and mending cables.

### Categorizing supercomputers

Before we can specify in meaningful ways the performance of computers, we really need a categorization which characterizes their basic features. In 1972 Flynn provided such a categorization. It has proved useful and has stood the test of time, partly because of its simplicity, but also because it concisely captures the key features of many different computer architectures. (It does have some deficiencies, but we shall ignore them, acknowledging only that there will be computers which do not fit easily into Flynn's scheme.)

SISD   *Single Instruction, Single Datastream* computers. This is the description of the typical, single-processor, standard or straightforward computer. It processes a single instruction at a time, and works on one piece of data at a time (see Fig. 12.3). It is akin to our single 'Jack of all trades' mending cables.

SIMD   *Single Instruction, Multiple Datastream* computers. This is the description of the typical, parallel array processor

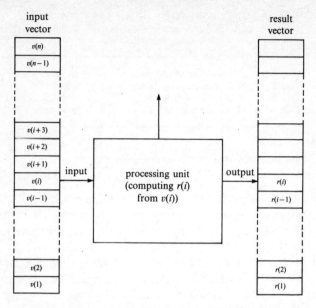

*Figure 12.3.* A schematic illustration of a SISD computer
processing a vector $v$ of data to compute the result vector $r$,
one element at a time, showing the $i$th stage of the
computation.

computer. The computer consists of a set of processors, all
of which are linked together so that they will all perform the
same instruction at any one time, but each works on its own
data. In a typical SIMD computer, each processor can
communicate with a number of other processors, typically
arranged by configuring the processors as an array (see
Fig. 12.4). This type of computer is akin to our set of four
'Jacks of all trades', all working in parallel and unison on
four cable-mending jobs. Examples of this type of computer
are the ICL DAP (Distributed Array Processor) and the
Burroughs Scientific Processor (BSP).

MIMD *Multiple Instruction, Multiple Datastream* computers. These
computing systems are arrangements of several instruction
processor units and several datastreams. The increasingly
common multiprocessor configurations are therefore
members of this class.

One of the first parallel features to be used to increase the performance of computers was pipelining, first exploited in a major way

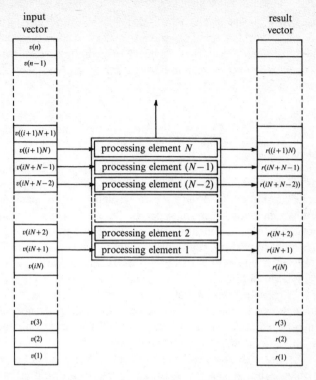

*Figure 12.4.* A schematic illustration of a SIMD computer, consisting of $N$ identical processors, processing a vector $v$ of data in groups of $N$ elements to compute the result vector $r$, at the time when the $N$ processors are processing elements $v(iN + 1)$ to $v((i + 1)N)$.

in machines like the IBM 360/91, CDC 6600 and ICL 2980. In pipelining, each operation is divided into a number of sub-operations, each of which is performed on a separate subprocessor, all of which can operate simultaneously. The division into sub-operations is an inherent feature of all computers; concurrent operation brought the performance improvement, as will be shown in the next section. Pipelining can be applied to arithmetic operations, as shown in the last section, or to 'administrative' functions such as instruction processing. This processing could consist of following the sequence: decode instruction, calculate addresses of data, initiate the fetching of data, execute operation on the processor, fetch the next instruction. In a pipelined instruction processing unit for this subdivision of operations, there would be a separate specialist subprocessor for each sub-operation, i.e. five subprocessors in all. They would work in parallel, and so, at any instant of time, five instructions would be at various stages of being processed through the pipeline. The accompanying performance improvement can be dramatic, as we shall now see.

### Measures of the performance of supercomputers

The basic measure of instructions per second (ips), or floating-point operations per second (flops), although adequate when considering only one computer (assuming it is a straightforward machine), is often not sufficiently sharp for supercomputers if we want to predict performance on different applications. Supercomputers are generally designed to perform the same task many times over on different data, or on a succession of data. For them, as with our cable-mending team of four, the performance can vary over a large range. If the four men are mending just one cable it takes four days (i.e. one per four days); if they have an infinite succession of cables, they can mend them at the rate of one per day (after an initial start-up time) – a fourfold performance improvement.

In 1981 Hockney and Jesshope found a better measure of expected performance than the single measure of peak computational rate. In our example of cable-mending, the difference between the peak performance and the performance on any particular problem was considerable. An important parameter that strongly influences the actual performance obtained on a given problem is the *vector length* – analogous to the number of cables to be mended in our example. Typically, for a pipeline processor and a vector of infinite length, the speed obtained is the peak performance. A meaningful measure is the length of the vector for which half the peak performance is achieved, because this gives a measure of how much replication is needed in the computing in order to achieve a high performance. Let us determine how these quantities relate to the actual machine characteristics and architecture.

In virtually all computers, any operation (such as an addition or a multiplication) is divided into a sequence of elemental operations. We saw in Table 12.2 how this was done for floating-point addition. Suppose an operation is divided into a sequence of $l$ steps. In addition, let us assume that each step takes one clock cycle to perform; let this time be $t$ seconds. Then to perform one operation in isolation takes $lt$ seconds, irrespective of the type of computer.

However, suppose now that we have a set of data vectors of length $n$ on which we want to perform this operation, say adding two vectors together. For example, one vector might represent the marks out of 50 for each of $n$ pupils in an applied mathematics class, and the second vector the marks out of 50 for those same pupils in pure mathematics, and we want to add the two marks together to form a percentage for each pupil.

### 'Standard' serial computers

For this type of computer, only one operation can be performed at a time, and so the time taken to perform the operation on the vectors of length $n$ is $nlt$ seconds. The rate at which results are produced is one every $lt$ seconds, and this is also the maximum rate, i.e.

$$r_{\infty,\text{serial}} = (lt)^{-1} \text{ results per second} \qquad (12.1)$$

The notation $r_\infty$ indicates this is the rate at which results are produced under the assumption that the vector is of infinite length. With a vector of infinite length, the computer is able to get up to its peak rate and any starting or stopping activities can be ignored. Such a rate is called the *asymptotic speed*, since it is the speed that would be approached as the vector length increased to infinity. As with our four men mending cables, the asymptotic rate of a cable mended every working day is the peak rate, but that is achieved only after an infinite number of days, because of the three days needed to start up all the team's activities. Notice that in such a computer system, each circuit is active for only $1/l$ of the time for each operation – just as each man in the cable-mending team works on each cable for only one day. As we saw, their performance could be improved by having them work on several cables at a time. This is the equivalent of pipelining in the computer.

### Pipelined processors

A processor with a pipelined arithmetic unit is one in which each of the sub-operations can be performed at one and the same time, but of course each must therefore be working on different data. To keep all the units busy, we must be feeding them a continuous stream of data, such as a vector of values (see Fig. 12.5). Suppose we feed a vector of length $n$. Then the time taken to perform the operation on all elements of the vector is

$$T = st + lt + (n - 1)t \qquad (12.2)$$

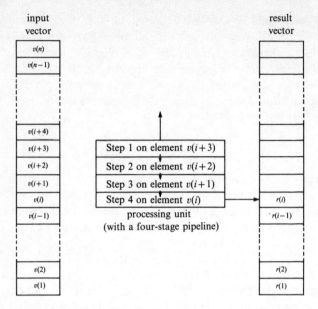

*Figure 12.5.* A schematic illustration of a pipelined
processing unit shown operating on a vector $v$ of data. The
processing unit is performing each of the four sub-operations
simultaneously, with each operating on successive elements
of the vector in turn, thus producing a result in the vector $r$
every time the sub-operation is completed (after the start-up
period).

where $s$ is the number of clock cycles needed to set up the pipeline
for this computation, and $lt$ is the time taken to perform the operation on the first element of the vector(s). By the time the operation
on the first element is complete, all but one of the sub-operations
needed for the second element will already have been performed
(namely $l - 1$ sub-operations) and all but two on the third element,
and so on through the pipe. During the next clock cycle, each
sub-operation will be performed on its next set of operands, producing the result for the second element of the vector, advancing each
of the other elements through the 'pipe' by one further operation,
and performing the first sub-operation on the new element which has
just entered the 'pipe'. Thus it will take $(n - 1)t$ clock cycles to
complete the operation on the whole vector.

The peak performance, given a vector of infinite length, is one
result every clock cycle, namely one every $t$ seconds, and so

$$r_{\infty,\text{pipeline}} = t^{-1} \qquad (12.3)$$

Thus a pipeline processor in which each operation is subdivided into $l$ sub-operations, and each sub-operation can be performed in parallel, has a maximum speed-up of $l$ times over its sequential counterpart.

The actual time taken to process a vector of length $n$ is given by

$$T = (s + l - 1)t + nt \qquad (12.4)$$

and we can rewrite this as

$$T = (n_{1/2} + n)t \quad \text{where } n_{1/2} = s + l - 1 \qquad (12.5)$$

Notice that if $n$, the length of the vector being processed, is $n_{1/2}$, then

$$T = 2nt \qquad (12.6)$$

that is, the time is twice that of the best possible performance. For this reason we call $n_{1/2}$ the *half-performance vector length* – it is the vector length required to achieve half the maximum performance.

The formula for the time taken to process a vector of length $n$ can be rewritten as

$$T = (n + n_{1/2})r_\infty^{-1} \qquad (12.7)$$

where $r_\infty$ is the maximum asymptotic performance. For typical supercomputers the performance rate $r_\infty$ is achieved with infinitely long vectors. The typical unit for $r_\infty$ is the Mflops.

It is important to recognize the significant difference between the two parameters $n_{1/2}$ and $r_\infty$ which appear in this formula. The maximum (asymptotic) performance, $r_\infty$, is primarily a characteristic of the technology used in the computer. Simply by changing the clock period, we can alter $r_\infty$ (e.g. halving the clock period doubles $r_\infty$). The value of $r_\infty$ does not reflect the architecture of the computer, namely how well that peak performance can be used. The half-performance vector length $n_{1/2}$ *does* reflect the computer architecture, in particular its impact on the user in terms of the amount of replication needed in a problem in order to exploit the performance capabilities. If a computer is designed to allow pipelining by dividing each operation into $l$ sub-operations, and permits each sub-operation to be performed in parallel (on different data values), this will affect $n_{1/2}$. Although the time taken for one full operation, such as an addition, remains unaltered, $r_\infty$ for each processor will remain the same. Thus $n_{1/2}$ is a useful 'one-parameter' measure of the architecture, in particular the parallelism, of a computer.

Looking now at the mathematics, there are two main points to be made. The first is that equation (12.7) has been obtained from a relatively simplistic model of a computer, what we call a 'first-order' description of the performance. If we assume that performance depends on the vector length $n$ of the data being processed, then we can postulate that the time $T$ taken to process that vector is related to a polynomial in $n$:

$$T = a_0 + a_1 n + a_2 n^2 + a_3 n^3 + \cdots \qquad (12.8)$$

The zeroth-order approximation is, then, simply a polynomial with highest term of order $n$ to the power zero, i.e.

$$T = a_0 \qquad (12.9)$$

which for our purposes is not very useful – it is saying that the time taken to process a vector of any length is constant, at $a_0$. A better *single-term* approximation for $T$ would be

$$T = a_1 n = r_\infty^{-1} n \qquad (12.10)$$

This is precisely the simplest measure we started with, namely the peak (asymptotic) performance. The first-order approximation, but including two terms, is a polynomial with highest term of order $n$ to the power 1:

$$T = a_0 + a_1 n \qquad (12.11)$$

and this is precisely the form of equation (12.7), with $a_0 = n_{1/2} r_\infty^{-1}$ and $a_1 = r_\infty^{-1}$. Higher-order approximations can be derived, but they are not very useful for this expression in this problem. Other features of the different computers are of more importance than getting better approximations that depend only on the vector length.

The second point to be made is that equation (12.7) has been derived by representing the computer as a model, and then analysing the features of that model. This contrasts with the alternative approach in which the results of examples are examined empirically, without considering the underlying process. The model-based approach provides a way of predicting the performance of computers, even before they are built, and hence focuses the attention of designers of computers on the main ways of improving performance. The same is true for the algorithm developers, software writers and users – a knowledge and understanding of the model helps them to develop and write computer code that will fully exploit the architecture of a particular computer, without having to be aware of the architecture itself, at least 'to first order'.

For a 'standard' serial computer, $n_{1/2} = 0$. For computers employing pipelining units, $n_{1/2}$ may vary for different arithmetic processes; it will depend, for example, on how an operation is subdivided and performed on the computer.

## Parallel processors

Another form of parallelism is that embodied in the SIMD class of processors – where a set of cooperating processors all perform the same operation simultaneously on a set of different data values. Suppose the computer has $N$ processors, so that it can process up to $N$ data items simultaneously. If the vector being processed is of length $n$, and $n \leqslant N$, then the time taken to do this operation on the $n$ values is independent of $n$, and is equal to the time $t$ for each processor to do just one of the operations. Since, for most problems where the supercomputer is really needed, $n$ is very large, let us concentrate on this case. If $n > N$, then we obtain an elapsed time, which is stepped:

$$1 \leqslant n \leqslant N, \quad T = t$$
$$N < n \leqslant 2N, \quad T = 2t$$
$$\begin{array}{c} \cdot \\ \cdot \\ \cdot \end{array} \qquad (12.12)$$
$$(k - 1)N < n \leqslant kN, \quad T = kt$$

We can now fit the time to process a vector of length $n$, assuming it to be given by the formula

$$T = (n + n_{1/2})r_\infty^{-1} \qquad (12.13)$$

This formula relates the time $T$ to the length $n$ of the vector. It is a linear relationship, of the form $y = mx + c$, where $m$ is the gradient. For equation (12.13) the gradient is $r_\infty^{-1}$. Relating this to the stepped relationship above, we find that the slope of the line which is the best fit to the stepped relationship (see Fig. 12.6) has $r_\infty = N/t$, and the intercept with the $T = 0$ axis is at $n = -\frac{1}{2}N$. Hence

$$r_\infty = N/t \quad \text{and} \quad n_{1/2} = \tfrac{1}{2}N \qquad (12.14)$$

Note that the peak performance for this type of computer may be obtained with a vector of length $N$, or an exact multiple of $N$. This contrasts with the pipelined architecture described earlier in which the peak performance is attained only on a vector of infinite length.

For a SIMD computer which uses pipelining within each processor it is straightforward to combine the arguments used to derive equations (12.4) and (12.13) with (12.14) to obtain the expressions

$$r_\infty = N/t \quad \text{and} \quad n_{1/2} = N(s + l - \tfrac{1}{2}) \qquad (12.15)$$

where $t$ is now the clock cycle of one sub-operation. It is instructive to examine these formulae. We see that by replicating a pipeline $N$

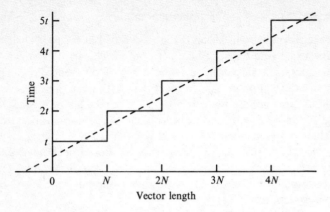

*Figure 12.6.* The time taken to process a vector of length $n$ on a SIMD computer with $N$ processors. The solid line is the actual time, the dashed line is the 'best-fit' linear (straight-line) approximation. The intercept at the point $-\frac{1}{2}N$ gives the vector half-length $n_{1/2} = \frac{1}{2}N$, and the slope gives $r_{\infty} = N/t$, where $t$ is the time taken for one operation by one of the processors.

times, we improve the (maximum) asymptotic performance by a factor of $N$, but we also increase the half-performance vector length by the same factor (see Fig. 12.7). Thus the total parallelism is increased $N$-fold, and we need this level of parallelism in the problem if we are to continue to exploit the computer's improved capabilities.

## MIMD computers

Is it possible – indeed meaningful – to represent the performance of the highly flexible architecture of a MIMD type of computer with the relatively simple first-order expression obtained above for other classes of computer? At first glance the answer is 'No'. A MIMD computer allows the different processors to be doing different instructions on different data – that is, independently. However, in solving a particular problem, if all the processors within the MIMD computer are to be contributing, there must be some cooperation at various stages of the computation. This process is called *synchronization*, and it is the cost of synchronization that is the penalty paid for the generality of the MIMD computer. Thus, if some of the processors finish their tasks earlier than others, and a synchronous process has to be performed before those processors can continue,

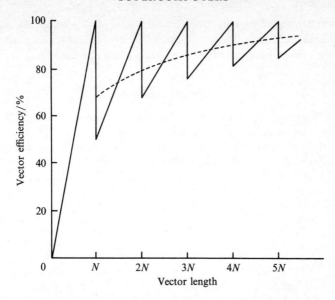

*Figure 12.7.* The efficiency of a SIMD parallel processing computer with $N$ processors. The solid line shows the actual efficiency of the group of $N$ processors when operating on a vector, which is 100% if the vector length is an exact multiple of $N$, and has a minimum when the vector length is one greater than an exact multiple of $N$. For this computer $n_{1/2} = \frac{1}{2}N$, and the approximate performance representation of equation (12.13) is shown by the dashed line.

the overall performance is degraded. The degradation in performance in equation (12.13) is represented by the parameter $n_{1/2}$, whereas the $r_\infty^{-1}$ term represents the best that can be achieved. These expressions, suitably interpreted, can provide a meaningful measure of the performance of the MIMD class of computers. If each processor has pipelining or parallel features itself, then these too can be naturally incorporated into the expression.

## Conclusion

Computers are already very complex entities; indeed, one of the key uses of existing computers is to design new computers! We cannot, therefore, even hope to represent all the features of complex computers, such as today's supercomputers, by simple models. Nevertheless, it is clear that a simple conceptual mathematical model of computer processing units can go a considerable way in characterizing key features of virtually all computers, at least for their performance on the typical problems for which they are to be used.

## Further reading

R. W. Hockney, 'Performance of parallel computers', in *High Speed Computation*, edited by J. S. Kowalik (Springer-Verlag, 1985).

R. W. Hockney and C. R. Jesshope, 1981, *Parallel Computers – Architecture, Programming and Algorithms* (Adam Hilger, 1981).

# THE CONTRIBUTORS

**Dr Julian Hunt** teaches and does research at the University of Cambridge in Trinity College and in the Department of Applied Mathematics and Theoretical Physics and the Department of Engineering. After graduating in engineering he did research at the Universities of Cambridge and Warwick, spent time in America and Africa, and worked for the Central Electricity Research Laboratories. He has worked mainly on fluid mechanics problems that arise in industry and the environment, for example flows in oil wells, electromagnetic stirring of molten metals, wind energy over hills, and dispersion of pollution. He is director of a small consultancy company that tackles environmental problems.

**Dr Nick Higham** is a lecturer in mathematics at the University of Manchester. He spent the academic year 1988/89 as a visiting assistant professor in the Department of Computer Science at Cornell University, New York. He obtained his PhD from the University of Manchester in 1985, and was awarded the international Alston S. Householder award for the best PhD thesis of 1984–87 in numerical linear algebra. His main research interests are in numerical analysis and matrix computations.

**Dr Andrew Noble** is chairman of Specialeyes plc, a fast-growing optical retailer. He graduated in mathematics at St Andrews, and in economics and statistics at Cambridge. He joined ICI in 1959 as an industrial statistician but then went on to hold a number of posts in operational research, planning and control. In 1973 he joined Debenhams, becoming a director in 1976 and a managing director in 1979. He has been a member of the Council of the IMA and was Vice President from 1986 to 1988.

**Dr John Rallison** lectures and researches in applied mathematics at Cambridge University; he is a fellow of Trinity College, where he was an undergraduate and then a postgraduate student. He spent a year working at the California Institute of Technology. His principal research interest is the mechanics of elastic liquids.

**Dr John Macqueen** is a mathematician working on power systems at the National Grid Research Centre (Leatherhead). He graduated in mathematics at St Andews in 1965, and for his PhD studied applications of the theory of generalized functions. In 1968 he joined the CEGB at its central research laboratories, where he worked as a mathematical modeller, with a spell of almost three years (1978–81) in London gaining valuable experience of the human factors in research. His research applications have included environmental issues, which stimulated his interest in the application of mathematics to biological systems.

**Dr Michael Thorne** is head of the School of Computer Studies and Mathematics at Sunderland Polytechnic. Apart from writing several books and contributions to learned journals, he is also active as a broadcaster and journalist. He has travelled extensively as a lecturer, both in Europe and further afield including Sri Lanka, the West Indies and the USA.

**Dr Jim Woodhouse** has been a lecturer in the Department of Engineering, Cambridge University, since 1985. His doctoral research was on problems arising from the vibrational behaviour of the violin; then followed a few years' work for a consultancy dealing with industrial noise and vibration. His interests cover a wide range of problems, all associated in one way or another with vibration, including the behaviour of musical instruments.

**Dr Nelson Stephens** is a senior lecturer in the Department of Computing Mathematics at the University College of Wales at Cardiff. He obtained his PhD in mathematics at the University of Manchester in 1965, and has taught at the University of East Anglia and at Pembroke College, Oxford. His research interests are cryptography, parallel computing, and the interrelation between computer science and pure mathematics, especially number theory.

**Dr David Jacobs** is New Technologies Manager with National Power. Previously he was branch manager of Power Plant

Performance Services at the CEGB's Central Electricity Research Laboratories, where he is responsible for much of the research into engineering, control, physics, chemistry, mathematics and computing. He studied mathematics at King's College, London, and then did research in applied mathematics at Imperial College, London. In 1970 he joined the CEGB research laboratories, spending his first nine years in the mathematics section pursuing a wide range of applied mathematics topics including fluid dynamics and electromagnetics, and making contributions in computational modelling and numerical analysis. He is a member of the Council of the IMA.

# FOR THE BEST IN PAPERBACKS, LOOK FOR THE 🐧

In every corner of the world, on every subject under the sun, Penguin represents quality and variety – the very best in publishing today.

For complete information about books available from Penguin – including Puffins, Penguin Classics and Arkana – and how to order them, write to us at the appropriate address below. Please note that for copyright reasons the selection of books varies from country to country.

**In the United Kingdom:** Please write to *Dept E.P., Penguin Books Ltd, Harmondsworth, Middlesex, UB7 0DA.*

If you have any difficulty in obtaining a title, please send your order with the correct money, plus ten per cent for postage and packaging, to *PO Box No 11, West Drayton, Middlesex*

**In the United States:** Please write to *Dept BA, Penguin, 299 Murray Hill Parkway, East Rutherford, New Jersey 07073*

**In Canada:** Please write to *Penguin Books Canada Ltd, 2801 John Street, Markham, Ontario L3R 1B4*

**In Australia:** Please write to the *Marketing Department, Penguin Books Australia Ltd, P.O. Box 257, Ringwood, Victoria 3134*

**In New Zealand:** Please write to the *Marketing Department, Penguin Books (NZ) Ltd, Private Bag, Takapuna, Auckland 9*

**In India:** Please write to *Penguin Overseas Ltd, 706 Eros Apartments, 56 Nehru Place, New Delhi, 110019*

**In the Netherlands:** Please write to *Penguin Books Netherlands B.V., Postbus 195, NL–1380AD Weesp*

**In West Germany:** Please write to *Penguin Books Ltd, Friedrichstrasse 10–12, D–6000 Frankfurt/Main 1*

**In Spain:** Please write to *Alhambra Longman S.A., Fernandez de la Hoz 9, E–28010 Madrid*

**In Italy:** Please write to *Penguin Italia s.r.l., Via Como 4, I-20096 Pioltello (Milano)*

**In France:** Please write to *Penguin Books Ltd, 39 Rue de Montmorency, F-75003 Paris*

**In Japan:** Please write to *Longman Penguin Japan Co Ltd, Yamaguchi Building, 2–12–9 Kanda Jimbocho, Chiyoda-Ku, Tokyo 101*

# FOR THE BEST IN PAPERBACKS, LOOK FOR THE 🐧

## PENGUIN SCIENCE AND MATHEMATICS

**The Panda's Thumb**   Stephen Jay Gould

More reflections on natural history from the author of *Ever Since Darwin*. 'A quirky and provocative exploration of the nature of evolution ... wonderfully entertaining' – *Sunday Telegraph*

**Gödel, Escher, Bach: An Eternal Golden Braid**   Douglas F. Hofstadter

'Every few decades an unknown author brings out a book of such depth, clarity, range, wit, beauty and originality that it is recognized at once as a major literary event' – Martin Gardner. 'Leaves you feeling you have had a first-class workout in the best mental gymnasium in town' – *New Statesman*

**The Double Helix**   James D. Watson

Watson's vivid and outspoken account of how he and Crick discovered the structure of DNA (and won themselves a Nobel Prize) – one of the greatest scientific achievements of the century.

**The Quantum World**   J. C. Polkinghorne

Quantum mechanics has revolutionized our views about the structure of the physical world – yet after more than fifty years it remains controversial. This 'delightful book' (*The Times Educational Supplement*) succeeds superbly in rendering an important and complex debate both clear and fascinating.

**Einstein's Universe**   Nigel Calder

'A valuable contribution to the demystification of relativity' – *Nature*

**Mathematical Circus**   Martin Gardner

A mind-bending collection of puzzles and paradoxes, games and diversions from the undisputed master of recreational mathematics.